The Water-Sustainable City

CITIES SERIES

Series Editor: John Rennie Short, *Department of Public Policy, University of Maryland, Baltimore County, USA*

As we move into a more urban future, cities are the main setting for social change, economic transformations, political challenges and ecological concerns.

This series aims to capture some of the excitement and challenges of understanding cities. It provides a forum for interdisciplinary and transdisciplinary scholarship. International in scope, it will embrace empirical and theoretical studies, comparative and case study approaches. The series will provide a discussion site and theoretical platform for cutting edge research by publishing innovative and high quality authored, co-authored and edited works at the frontier of contemporary urban scholarship.

Titles in the series include:

Cities as Political Objects
Historical Evolution, Analytical Categorisations and Institutional Challenges of Metropolitanisation
Edited by Alistair Cole and Renaud Payre

The Water-Sustainable City
Science, Policy and Practice
David Lewis Feldman

The Water-Sustainable City

Science, Policy and Practice

David Lewis Feldman

Professor, Department of Planning, Policy and Design and Professor of Political Science, University of California, Irvine, USA

with contributing authors

Stanley B. Grant

Department of Civil and Environmental Engineering, Henry Samueli School of Engineering, University of California, Irvine, USA

Ashmita Sengupta

Southern California Coastal Water Research Project, USA

Lindsey Stuvick

Irvine Ranch Water District, USA

Meenakshi Arora

University of Melbourne, Australia

Vincent Pettigrove

University of Melbourne, Australia

Kristal Burry

Public Interest Advocacy Center, Ltd, Australia

CITIES SERIES

Cheltenham, UK • Northampton, MA, USA

Published by
Edward Elgar Publishing Limited
The Lypiatts
15 Lansdown Road
Cheltenham
Glos GL50 2JA
UK

Edward Elgar Publishing, Inc.
William Pratt House
9 Dewey Court
Northampton
Massachusetts 01060
USA

Paperback edition 2018

A catalogue record for this book
is available from the British Library

Library of Congress Control Number: 2016949956

This book is available electronically in the **Elgar**online
Social and Political Science subject collection
DOI 10.4337/9781783478576

ISBN 978 1 78347 855 2 (cased)
ISBN 978 1 78347 857 6 (eBook)
ISBN 978 1 78347 856 9 (paperback)

Typeset by Servis Filmsetting Ltd, Stockport, Cheshire
Printed and bound by CPI Group (UK) Ltd, Croydon, CR0 4YY

Contents

Preface

Throughout history cities and water have had what amounts to a kind of "love–hate" relationship. To have cities, one must have reliable and abundant freshwater sources. However, where we choose to build cities – and what cities become after they grow – can place enormous strain on locally available water resources. Growth and development can also severely degrade the quality of water consumed by other users, as well as the condition of the surrounding natural environment. As a result, since the inception of the earliest cities, urban governments have embarked on one of two principal strategies to compensate for these challenges: they find ways to bring water from considerable distance to quench their population's thirst or – to paraphrase one of the major characters from the movie *Chinatown* – they annex neighboring areas with freshwater in order to acquire rights, and access, to additional supplies.[1]

By the late twentieth century, another set of options began to emerge – applying various technical "fixes" to enhance local supply, ranging from desalination to stormwater harvesting, water banking, improving end-use efficiencies through installing low-flow appliances, and reuse of wastewater. While these remedies can make urban life better, and provide ample, clean, potable water to residents, they are not "cost-free." In some instances (for example, desalination) they may be economically expensive and provide potable water supplies costing far more than conventional sources. Moreover, while generalizations are not always reliable indicators, it would appear that the far larger and more complex a proposed solution to these problems becomes (for example, large water recycling facilities), the greater the likelihood that other environmental and public acceptance challenges may arise – including the safety and security of the water produced and the perception that the remedy or solution is not fair and equitable, and will not "fit" into an urban environment without undue adverse impacts to aesthetics, noise pollution, neighborhood character, and a host of other issues.

In short, adoption of these technologies may pose political and social conundrums at least as dramatic as the problems they are designed to solve – and that's what this book is intended to address: the economic, legal, institutional, environmental, and political challenges generated by efforts to establish water-sensitive cities, including the barriers that must be overcome in transitioning to a more adaptive or water-sensitive approach.

A water-sustainable city is one that adopts – successfully – these and other technologies to support growth and development. However, we contend that to become truly water sustainable requires that established technologies are supplemented by low-energy – and low carbon footprint – approaches such as biofilters for extracting contaminants and recharging groundwater basins and aggressive conservation measures to reduce demand. Both may also impose fewer adverse effects on the environment, thereby improving water productivity, lowering energy (and carbon) footprints, and improving human and ecosystem health.

Low-energy, low carbon footprint approaches go by many names, reflecting their diversity in size, scale, cost, and application. What they all share in common, however, is an emphasis on resilience: ensuring that under adverse as well as "normal" conditions, cities will have a dependable, reliable, safe and sufficient supply of clean and usable water. Put another way, a resilient city, from the standpoint of water, is one that is able to reduce its vulnerability to harm from supply disruptions from various hazards, including climate change (Padowski et al., 2015). To achieve this goal, these alternative approaches share another feature in common – one that distinguishes their philosophy and method of implementation from more "conventional," large-scale approaches. The critical feature of these so-called resilient water systems is to closely integrate water management to land-use and urban planning. In part, they seek to do this by fusing centralized and decentralized systems for water supply and treatment, and they emphasize diversity of supply sources throughout entire urban watersheds (Daigger, 2009; 2011; Pahl-Wostl et al., 2007; Gersonius et al., 2013; Gleick, 2003).

Such approaches in the United States have been referred to as low-impact development (LID). In the United Kingdom, they go by the name Integrated Urban Water Management (IUWM). And in Australia, the terms Water-Sensitive City (WSC) and Water-Sensitive Urban Design (WSUD) have become popular labels. A variety of naming conventions continue to be debated in the water community, in part reflecting the evolution of philosophies over urban drainage and other issues of water management (Brown et al., 2009; Wong et al., 2011; Fletcher et al., 2013a).

To investigate these issues, this book focuses on two interrelated issues of critical importance to cities and the environment. The first of these is how our rapidly urbanizing planet impacts global water availability and quality. The second is the ways in which various threats to freshwater – including global climate change and pollution – impact water infrastructure, public health, and societal welfare in cities. We contend that while the world's freshwater faces enormous pressures from drought, weather variability, and unceasing demands for more food and energy, large cities generally – and megacities in particular – compound these stressors in two distinct ways.

First, cities are often located some distance from the water sources needed to maintain their teeming populations, compelling them to divert water from outlying rural areas. These areas, in turn, often produce the food and fiber needed to support the former. And second, soaring birth rates, in-migration (the latter often from these same outlying areas which are suffering from water and food shortages), and growing suburban sprawl, place extra burdens upon water infrastructure – exacerbating health and hygiene problems, and worsening risks from sea-level rise and flooding.

While the phenomenal growth of megacities, a trend which began in the late twentieth century, has become a defining nexus for numerous environmental problems including air pollution, growing greenhouse gas emissions and natural habitat destruction until recently, the effects of urbanization on the world's freshwater have been little studied. These effects are worthy of our attention for two reasons. First, megacities are increasingly contributing to water conflicts between themselves and their outlying regions and, by implication, worsening water disputes among neighboring countries. And second, while the magnitude of these problems is severe, the problems themselves are not really new – many are traceable to cities in antiquity. One innovation of this book is that we seek to link current problems to these historical antecedents.

Our thesis is that global climate change and population growth demand creative, low-energy, multi-disciplinary, and multi-benefit approaches to sustaining water resources. This thesis draws upon the work of the principal author and his collaborators through a bi-national project funded by the National Science Foundation. This project, the Partnerships for International Research and Education (PIRE), is a joint effort of investigators at UCI, UCLA, UC San Diego, and the Southern California Coastal Waters Research Project (SCCWRP) in the U.S., and scientists at the University of Melbourne and Monash University in Australia. Over the past three years, we have undertaken efforts under this project to catalyze, through research and education, the development and deployment of low-impact alternatives for improving water productivity while protecting human and ecosystem health.

Not only does this project link five different universities in two water-stressed regions of the world (the southwest U.S. and southeast Australia), but our Australian partners are world leaders in these topics, as evidenced by the world's first distributed application of these technologies in an urban catchment (Little Stringybark Creek), a university–industry Cooperative Research Center on water-sensitive urban design valued at $117M, the world's largest Wastewater Stabilization Pond, and a world-class facility for visualizing interfacial momentum and mass transport. By facilitating joint research and knowledge sharing, the PIRE has accelerated education and training in this critical area of water sustainability, and has diffused knowledge about sustainability options to U.S. middle-school and high-school students, undergraduate STEM majors, graduate students, post-doctoral researchers, and practitioners.

Part I of this book comprises four chapters. Chapter 1 examines our uncertain water future and considers the possibility of a water-sustainable city and how that city might differ from contemporary metropolises in regard to its use and management of water. Chapters 2, 3, and 4 respectively consider some major lessons from an urban ecology of water – a historical view of urban water management; the roles of civil engineering, law and institutions in urban water management; and different approaches that have been adopted – both traditional as well as contemporary – in managing urban water problems.

Part II explores the water–energy footprint of some large cities and how urban form can impact water use. This section considers two interrelated issues: (1) the relationship between water and energy in large cities (what is sometimes referred to as the water–energy nexus); and (2) how cities value water – an emphasis on the economic valuation of water use. Chapter 5 looks at the water–energy nexus through the lessons of the PIRE project especially, while Chapter 6 discusses ways in which the value of water can be measured, as well as economic and non-economic tools for valuing and managing it – including how to induce conservation. Other economic tools available to manage urban water more efficiently are also discussed.

Part III proposes a way forward, with a dual focus on technology and policy. Chapter 7 examines the urban stream syndrome and how it can be alleviated – an especially important topic in the contemporary city with respect to managing both ecological and societal needs for water. Chapter 8 discusses so-called low-energy and, thus, low-carbon options for water supply, along with innovations in infrastructure. These approaches are all critical to meeting the challenge of better using available water resources in cities. Chapter 9 examines the future of water governance and management for achieving a water-sensitive city. Finally, we will conclude (Chapter 10) with some thoughts regarding future research needs for decision-makers.

NOTE

1. The character, played by John Huston, stated: "if you can't bring the water to Los Angeles, then you have to bring Los Angeles to the water," referring to plans to acquire the San Fernando Valley and annex it to the city. Such a plan, in less fictitious form, occurred in the early twentieth century when Los Angeles acquired the water rights to the Owens Valley and built an aqueduct to divert water to the city some 250 miles distant.

Acknowledgments

This book was partially supported by a grant from the U.S. National Science Foundation Partnerships for International Research and Education (OISE-1243543).

I wish to add special thanks to two former UCI students, Michael Sahimi and Neeta Bijoor, for their contributions to the book. Although neither one was properly a part of the PIRE project from which much of the book's ideas stem, both undertook important research on water conservation projects while at UCI and their analyses and findings are featured in Chapters 6 and 8, respectively.

PART I

Our Uncertain Water Future, Our Precarious Water Past

1. Introduction: what would a water-sustainable city look like?

This book explores the dynamic relationship between cities and water. It is the product of insights acquired through our research on the PIRE project, which is focused on developing and deploying low-energy options for improving water productivity while taking into account the environmental and social values of water. It is designed for advanced undergraduate and graduate students, as well as for water professionals, policymakers, and members of the public who aspire to a more in-depth understanding of contemporary water issues.

This book focuses on two subjects. First, how are fast-growing cities, both in developing and developed countries, impacting freshwater availability and quality in their region of influence? Second, what tools for analyzing the environmental and social impacts of urban water consumption – and for choosing approaches and policies for addressing various threats to freshwater management – are available? As noted in the Preface, our goal is to help cities manage water while confronting the challenges of climate change, maintaining ecosystem form and function, and reducing the water and energy footprint of cities so as to put them on a more sustainable path that quenches our thirst for water while preserving ecological health and societal well-being.

Human overuse of water resources has placed enormous pressures on freshwater availability. Cities – especially large ones – exacerbate these problems in three distinct ways. First, they are often located some distance from the water sources needed by their populations. In the past, at least, this has compelled them to build new infrastructure to divert water from increasingly distant outlying rural areas, thus disrupting their social fabric and their environment. More recently, cities have been turning to local and lower-quality sources of water, largely because the distant high quality sources have been tapped out. This is not a new problem: as we will show in Chapter 2, it is as old as antiquity.

Second, increasing urbanization due to population growth, economic change and sprawl places huge burdens upon the institutions, as well as the infrastructure, that deliver and treat urban water, especially in developing countries where these challenges are compounded by rapid growth, weak institutions, and insufficient funding capacity. While greater concentration of people in cities may lower unit costs for many forms of water infrastructure, the need to expand water supply and treatment networks over vast areas may increase the likelihood of distribution system leaks and other failures (United Nations Human Settlements Programme, 2011).

And third, the *spatial* "footprint" caused by sprawling horizontal urban development and annexation imposes numerous water-related problems. A few of the most important of these include paving of city streets and commercial districts (which contributes to pollutant runoff and diminished groundwater recharge); consumption of water for parks and outdoor residential use (increasing evapotranspiration and taxing local supplies); increased urban "hot spots" due to the alteration of land cover; and, urban waste discharges that affect local to global biogeochemical cycles and climate (Grimm et al., 2008). While cities themselves present numerous water problems, they also harbor many of the solutions to the sustainability challenges of an urbanizing world.

This book will focus on innovative solutions that foster water conservation, repurposing of water, wastewater reuse, and more generally: local treatment techniques that require less energy, rely on natural processes, and generate environmental value (such as constructed wetlands), as opposed to established practices such as long-distance diversion and large, centralized, energy intensive treatment and distribution approaches. After providing an overview of the complex water challenges facing fast-growing and established urban areas if we stay the course, we will strive to understand how cities have tackled their water challenges based on examples from the past as well as the present.

THE FUTURE OF CITIES AND WATER: THE PRESENT AS PROLOGUE

What might cities of the future look like if they could use water more sustainably – in ways that protect environmental quality, promote economic development, and foster just and equitable resource allocation and management? This chapter provides an overview of some of the principles and practices that might characterize such a future city, particularly drawing upon methods to regenerate, conserve, and substitute for freshwater, as well as impediments to adopting these methods – especially as seen in Australia and the U.S. A good place to begin might be to examine cities and their water uses in the present.

As of 2016, more than half of the world's population lives in cities – some 3.5 billion souls. Putting this into further perspective, at the current rate of population growth, the world's urban population is growing by some two people every second. And, in developing countries, this growth rate is the equivalent of some five million new urban dwellers every month. Moreover, while there are many reasons people choose to live in cities, even in less developed nations where water quality tends to be sub-standard (a topic we take up later in the book), in general, water conditions in cities tend to be far better than in non-urban areas. According to the United Nations, as of 2010, 96 percent of the world's urban residents had access to improved sources of drinking (as compared to 81 percent of those living in rural areas). For sanitation facilities, the disparities are even more profound: 79 percent of the world's urban dwellers had access to improved sanitation facilities while only 47 percent of rural dwellers did (Swyngedouw, 2007).

"Megacities" composed of tens-of-millions of people comprise a phenomenon that is especially acute in developing countries. Some 80 percent of the world's city dwellers live in developing countries. While spanning various levels of modernity, nevertheless, more than two-thirds of the world's urban residents live in cities in Africa, Asia, and Latin America, where development lags behind most of Europe, North America, and other industrialized countries. Since 1950, the urban populace of these regions has grown five-fold, while in Africa and Asia alone, urban populations are expected to double by 2030 (Satterthwaite, 2000).

Perhaps more compelling than these raw statistics, however, is that despite the progress made in improving water provisioning and sanitation in the world's cities, these services face a race against time and place. Over the next 35 years or so, the world's urban population is expected to increase from some 4 billion to 6.3 billion by 2050 – a higher rate of growth (36 percent) than the overall projected rate of the world's population (26 percent) during this period (UN Water, 2014). Most of this growth will occur in cities in the poorest of the world's nations, where densely populated slums replete with sub-standard housing already pose enormous challenges regarding access to safe water and sanitation.

In part because of the rapid rate of urban growth in the poorest nations of the world, between 2000 and 2008 urban areas worldwide witnessed a deterioration in both water and sanitation coverage, with access to sanitary toilets decreasing by some 20 percent (UN Water, 2014). Moreover, urban settlement is the main source of point-source water discharges such as raw sewage.

The same United Nations reports that cite these dire statistics also pin their hopes for improvement on the very kinds of initiatives that we identify as characteristic of a water-sensitive city – a reliance not just upon sophisticated and centrally managed technologies, but on exploitation of comprehensive urban water planning, investment and associated operations in the cities of the future. In essence, developing countries may be able to "skip" the mistakes made by more developed nations and develop distributed and eco-sensitive approaches to stormwater management and other approaches more quickly than developed countries.

With this challenge as context, our research through PIRE, and that of other investigators, defines a water-sustainable city as a place that imposes minimal adverse impact on water resources; pursues options for the delivery, treatment, use, and management of water that embrace a positive water–energy nexus – that is, high water productivity with low energy use (and thus, a low carbon footprint); and embraces these initiatives as parts of a "holistic framework" embracing all aspects of the water system in a city (Howe et al., 2011). Such holistic development revolves around adoption of three principal sets of options (Figure 1.1) that include low-energy approaches for substituting, regenerating, and conserving freshwater.

Substitution is the use of lower-quality freshwater for urban uses such as industrial cooling and landscape irrigation. At the household level, it could encompass watering gardens with rainwater from a tank, as well as flushing toilets and washing laundry with treated stormwater effluent from a rain garden or similar structure. *Regeneration* involves deriving high-quality water from lower-quality water, including wastewater and stormwater through direct or indirect (for example, treatment and groundwater replenishment) methods. As shown, a waste stabilization pond (WSP) can transform household sewage into high-quality water for irrigating an orchard. Finally, *conservation* includes reductions in use through water-saving appliances (for example, drip irrigation, dual-flush toilets, low-flow shower roses, front-loading clothes washers), as well as by use of innovative rate-structures, and even water-less toilets and urinals – as well as vigilant repair of leaks in distribution systems. Related infrastructure includes conventional drinking water treatment plant (DWTP); wastewater treatment plant (WWTP); and river diversion (supplying the orchard).

Keys to these innovations' acceptability and likelihood of adoption include, first, changes in public confidence and perceived competence in those responsible for managing these innovations brought about through public education to hasten trust (Po et al., 2005). In effect, cities of the future must undergo a "paradigm shift" to overcome bureaucratic inertia, developer concerns with returns-on-investment, tendencies toward fragmented urban planning, and resistance to public–private partnerships in water governance decisions (Saha and Paterson, 2008; Van de Meene et al., 2011). Second, policy – particularly regulatory – change also will be needed to establish scientifically supportable risk-criteria; enhance public confidence in water reuse; and eliminate concerns about inconsistent standards between different jurisdictions (Nellor and Larson, 2010). Third, as is graphically shown in Figure 1.1, these innovations are best

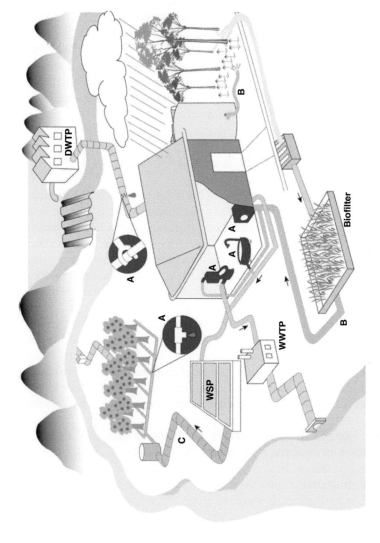

Note: (A) substitution; (B) regeneration; (C) reduction.

Source: Grant et al. (2012).

Figure 1.1   Toward a water-sensitive city of the future

7

introduced as distributed household or neighborhood-level improvements, or "add-ons" to existing water infrastructure. This makes them more practical and adaptive in their implementation because they are not being marketed as replacements for traditional infrastructure, which has the confidence of the public due to its proven reliability.

But why is a water-sensitive city really needed? Why won't traditional approaches do? To understand the justification for a water-sensitive city, we have to consider two other factors: climate variability/climate change as a set of variables affecting water availability, particularly in regions already water stressed; and the quest for meeting future water needs in a manner that is environmentally sustainable – a demand that has become especially tractable in the world's most highly advanced economies.

CLIMATE CHANGE, WATER AND CITIES

In recent years a strong consensus has grown among climate scientists, hydrologists, and others who contend that continued carbon dioxide and methane emissions will alter the climate in various ways. Global average temperatures will rise, for example, and dramatic changes in precipitation could adversely affect the world's freshwater supply. Rainfall intensity may increase in some regions and decline in others, while the seasonal balance between snow and rain might also shift, affecting local economies. Higher temperatures could increase evaporation and *transpiration* – the rate at which plants give up moisture to the atmosphere – while reduced soil moisture affects farming. Over longer time periods, shifts in vegetation cover over entire regions – from forest to grassland or grassland to desert – may occur. Accelerated melting of polar and glacial ice, another probable result of climate change, would lead to greater sea-level rise and saltwater intrusion into coastal estuaries, affecting fisheries and threatening urban drinking water supplies, and there are several positive feedback loops that could accelerate and exacerbate climate change, for example release of methane to the atmosphere as ice sheets and permafrost melt (IPCC, 2014).

Many scientists believe these changes are not only likely scenarios, but that current protracted drought in some regions, and unprecedented flooding in others, are harbingers of worse to come. While debate over whether and to what extent any given climatic event may be attributable to global climate change is far from settled, there is growing agreement that the increasing frequency of water-related extreme climate events is probably the result of human-induced climate change, including periods of more frequent and severe drought (Gleick and Heberger, 2012: 1–14; Griffin and Anchukaltis, 2014).

Table 1.1   Some adaptation options for freshwater

Supply side	Demand side
Prospecting and extraction of groundwater	Improvement of water-use efficiency by recycling water
Increasing storage capacity by building reservoirs and dams	Reduction in water demand for irrigation by changing the cropping calendar, crop mix, irrigation method and area planted
Desalination of seawater	Reduction in water demand for irrigation by importing agricultural products, i.e., virtual water
Expansion of rainwater storage	Promotion of indigenous practices for sustainable water use
Removal of invasive non-native vegetation from riparian areas	Expanded use of water markets to reallocate water to highly valued uses
Water transfer	Expanded use of economic incentives including metering and pricing to encourage water conservation

Source: IPCC (2007), Table 3.5. Also, see IPCC (2014), p. 27.

What does this portend for cities? Perhaps most critically, it means a growing reliance upon adaptation measures that can enhance our ability to manage water supply while hopefully attenuating urban water demands in the face of climate uncertainty. This requires imaginative management as well as good science, and it further depends on the ability to translate knowledge into language useful to decision-makers and the public. Moreover, activities undertaken for reasons other than climate change will be needed – what are sometimes referred to as "no regrets" strategies that have numerous benefits, such as the ability to accommodate population growth and growing demands – even if temperature or precipitation does not vary significantly over time (see Table 1.1).

As regards the aspiration to meet growing urban water needs in an environmentally sustainable manner, concerns with climate change and population growth have definitely affected the inclination of water resource planners and others to embrace new methods for providing water supply which avert – whenever possible – needs for large-scale, centralized and expensive to maintain infrastructure. Alternatively, newer, more resilient, and more "self-reliant" methods have arisen – self-reliant in the sense that cities can obtain reliable water supplies by better harvesting of local sources without resort to distant diversions or transfers of water across basins. Such remedies as we shall see include the harvesting, storage, and reuse of stormwater; the biofiltration of mildly polluted runoff; and the reuse/recycling of various grades of wastewater for different purposes.

OTHER MOTIVES FOR A WATER-SUSTAINABLE CITY: EMBRACING GROWTH, ASSURING LIVABILITY

While concern over climate change is a major impetus for adopting sustainability, it is by no means the only motivation. The International Water Association in its 2013 *Cities of the Future* report suggests adoption of these approaches should be undertaken in order to ensure that cities are both livable and resilient in the face of a growing urban population. Moreover, it suggests, these approaches should be adopted through methods that: engage various members of the public; ensure collaborative implementation; promote localized and adaptive solutions that can also be disseminated to other urban sectors; and promote "water literacy" among the urban public. Finally, IWA recommends that "all water is good water" and that future efficiency must "match quality to use" (International Water Association, 2013).

The IWA approach requires inter-disciplinary methods to be applied to understanding – and solving – urban water problems. Similar conclusions and recommendations have been reached by the European Union, and UNESCO (whose recent efforts on behalf of water-sustainable "cities of the future" promote interdisciplinary research from a long-term perspective). The International Water Association's "cities of the future" program, which connects water professionals from over 130 countries on many of these issues, and which recently issued a Declaration on cities of the future, has no doubt abetted this global diffusion of ideas (Howe et al., 2011: 30; UNESCO-IHE, 2011; International Water Association, 2013).

Our discussion in the following chapters weighs technological obstacles that must be overcome, as well as the cultural, social, economic, political, and legal hurdles that must be surmounted in order to achieve these goals. A theme of this book is that integrated urban water management is an approach to water that considers the management of demand a source of water, as much the augmentation of supply – and that water quality and quantity must be conjointly managed.

One major conclusion we will draw throughout this book is that water should be managed in cities to achieve – as close as possible – a "closed loop" system. In other words, what we should aspire to is not once-through water supply, stormwater, and/or wastewater streams, but a renewable, reusable resource (see Figure 1.2). How we can approximate this aspiration through better applying our knowledge to the realities of city life is the

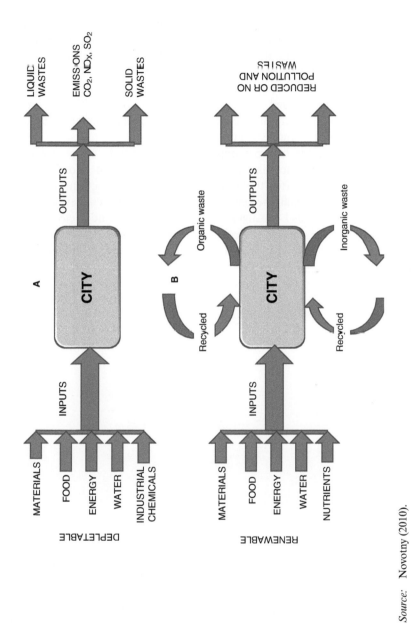

Source: Novotny (2010).

Figure 1.2   The concept of a "closed loop" sustainable city

subject of the following chapters. We begin, in Chapter 2, by discussing how cities have historically thought about these issues in different times and places.

In looking toward policy solutions, we will discuss two emerging issues. First, the precise urban form that growing megacities take can have a profound influence on water use and water supply impacts. For instance, sprawling horizontal urban development imposes numerous problems, including paving of streets and commercial districts, and diminishing of groundwater recharge (see Chapters 5, 7, and 8, especially). And second, many megacities have learned – in sometimes impromptu and improvisational ways – how to mitigate the pressures of growth and climate variability on water stress (for example, Chapters 3, 4, and 7).

Improvisations include better managing water demands, coupled with compensatory efforts for low-income groups who can least afford these innovations. We discuss how these steps can lead to reductions in per capita water use and forestall the need for additional supplies in developed-country cities as diverse as Los Angeles, New York, Tokyo, and Melbourne. While not a megacity, the latter's environmental footprint upon its surrounding region – particularly in light of Australia's recent Millennium Drought – is comparable to that of one.

2. Lessons for an urban ecology of water: historical views, environmental experiences

The very existence of cities requires a supply of water from surrounding environs so as to enable a large concentration of residents to settle in a single place. What can we learn about an urban ecology of water from past experience? This chapter examines urban water management in cities beginning with antiquity, starting in the Mediterranean world, in an effort to extract major lessons for achieving a water-sustainable city. These lessons may include learning from failures or controversial efforts to enhance local supplies.

Historically, cities emerged as centers of commerce and trade when rural food production advanced to a stage sufficient to allow people to congregate in smaller, more hospitable places. Food production, of course, was reliant on a comparably sufficient supply of harnessed water – and cities themselves likewise awaited the ability to provide such supplies as a means of supporting larger and larger populations. Throughout history, cities have sought reliable, safe, and plentiful supplies through infrastructural, economic, legal, and political strategies (for example, Jansen, 2000; Kamash, 2012; Freeman, 2004; Shaughnessy, 2000).

While Rome, for instance, was highly regarded for the sophisticated engineering of its elaborate aqueduct system, we know that this system was designed not just for water provision to cities, including Rome, but was equipped with various "branch lines" to provide ample supply to rural, agricultural lands supporting these cities – especially in the Eastern Mediterranean (Kamash, 2012). This fact further affirms the longstanding connection between cities and their outlying regions with respect to water management.

Another generalizable phenomenon with regard to cities and water is that they have long sought ways to control their local water sources and they also have struggled to harness the means to develop infrastructure to protect the safety, potability and cleanliness of household water, as well as to control against threats to the built environment such as flooding.

In the ancient Near East, for instance, cities emerged as centers of innovation for the development of waterworks to distribute freshwater and dispose of sewage and waste, as well as the legal and policy innovations introduced for water provision in the ancient city. This was true, for instance, in the Tigris–Euphrates valley, Assyria, Nile valley, and Jericho. Later, as large civilizations dominated by Roman-Byzantine hegemony arose, a major preoccupation of urban political elites was assuring adequate water to support growth.

This pattern, while characteristic of cities in antiquity (for example, Rome, where aristocratic classes or municipal agents appointed by the emperor expanded aqueducts beginning in the first century; McElwain and Herschel, 1925; Heather, 2006), tended to be shaped in one of two ways. State control was either directly exercised by an elite, or waterworks' installation and maintenance was managed and controlled by representatives of the population through free local management of hydraulic installations under the collective authority of the "hydraulic community" (Braemer et al., 2010).

In Europe, this approach continued in emerging cities after the fall of the Western Empire, despite the fact that many of the civil engineering marvels of antiquity were – for a time – lost or forgotten. Another was managing conflict over water supply (for example, Nile basin settlements in the third millennium BC) – a perennial issue even today in cities such as Los Angeles, New York, Mumbai and Mexico City. Historical antecedents illuminate how aspirations for growth and prosperity affect water management, place pressures on quality, and lead to incessant demands for more water – legacies which have endured in urban water management.

In thinking about how, and why, cities use water the way they do, it is useful to conceive of urban water management as revolving around three major issues: (1) the aspiration to harness and control local water sources; (2) the need to address equity (assuring that the provision of water serves all those living in a city, even if their power over water decisions varies); and (3) the importance of long-term aspirations for urban growth and development upon governance – seen through infrastructure development. We examine three cities from the Mediterranean region – an area that experienced dramatic urban growth in ancient times – to explore these issues.

ANCIENT CITIES AND CONTROL OF WATER SOURCES

In antiquity, urban elites often favored the making of water decisions without consulting interests outside of cities. Fortunately, and as a practical matter, there were not usually very many people living outside of the urban area who needed to be consulted. This was the prevailing pattern in Rome, for example, where aristocratic classes or municipal agents appointed by the emperor expanded the system of aqueducts beginning in the first century, and employed slave labor to build them. Variations of this pattern continued in European cities after the Western Empire's fall. In our era, exclusivity of decision-making – as we will see – is made more problematic by climate change and population growth, especially in developing countries lacking capacity to improve water infrastructure.

It is important to note at this juncture that prolonged drought and climate variability has been a well-documented problem in ancient civilizations, generally. We now know, for instance, that the collapse and abandonment of the Central Maya lowlands in the Yucatán Peninsula of present-day Mexico were the result of "complex human–environment interactions," landscape alterations, and other high-stress conditions made worse by prolonged aridity (Turner and Sabloff, 2012).

Water decision-making in ancient Athens, Rome, and Barcelona exemplifies some of these decision-making patterns in somewhat contrasting ways. In early Athens, homes and settlements were originally self-supplied, drawing extra water, when needed, from public supplied fountains built by local aristocrats and kings. Pisistratus (*c.* 560 BC) – and his sons – who seized power and became dictator, organized an effort to formally bring water into the city through rock-cut conduits and pipes from hills north of Athens. Elaborate fire clay pipes with round inspection holes were connected to basin-reservoirs that served both as fountains and as places where residents could draw potable water – water supply junctures provided by public officials.

Flood control works were also built at this time – stone-lined drainage channels, the first of which was probably built in the fifth century BC to convey rainwater into local streams. These works conveyed water draining from the slopes of the Acropolis, Areopagus, and Pnyx (Connolly and Dodge, 1998: 10, 14). Public waste provision in Athens came much later and dates to the period of Roman rule in the city. Officials who displayed special managerial skill were elected to supervise water supplies and public fountains in Athens (Hughes, 2014: 175).

Rome's legacy of water supply provision is by far the best known and chronicled of any ancient city. This is due, in part, to the fact that much of the city's water infrastructure is itself well preserved. However, it is also the result of the political importance accorded to the provision of water, attested to by the writings of Sextus Frontinus, the city's curator of aqueducts during a portion of the reigns of the emperors Flavius and Trajan, and a prolific author. After construction, local governments were obliged to maintain and repair water supply systems, not only in Rome itself but throughout the empire. Beginning with Augustus, special magistrates (aquaria) were appointed to oversee these activities (Purcell, 1994: 161; Garnsey and Saller, 1987: 33; Hughes, 2014: 175).

The first urban Roman aqueduct, the Aqua Appia, was built in 312 BC and carried water from springs 12 km east of the city. At the height of its power (*c.* second century AD) Rome had a total of eleven aqueducts, nine of which existed at the time of Frontinus' writings, with a total capacity of nearly 1 million cubic meters per day or approximately one thousand liters of water per person per day (Coarelli, 2007: 447–9). The Roman settlement of Barcino (later, Barcelona) founded in the late first century BC, built an aqueduct from the Besos River through the early walled-city, running along a Roman road north into a gorge leading to the Valles territory (Museu d'Historia de Barcelona, 2012). Upon entering the settlement, the state-built aqueduct distributed potable water to a series of "castella aquae," or tanks, to fountains, spas, and other distribution points scattered throughout the settlement. As in other parts of the empire, once Rome fell, maintenance of Barcino's aqueduct declined, and while the Besos aqueduct continued in use for over a thousand years, its maintenance and repair fell into neglect and local supplies became increasingly reliant on wells and cisterns. For all three of these cities, two compelling lessons emerge. First, access to external regional sources of water was perceived as absolutely necessary to ensure stable settlement and development. Second, it was a governmental responsibility to ensure that this need was authoritatively met. As far as is known, little opposition was demonstrated by outside forces, probably because the cities and their public works were guarded and usually fortified, and because few potential opponents lived outside city settlements to oppose these plans.

Equity and Fairness in Water in the Ancient City

Historically, urban water management has been characterized by multiple decisions regarding distribution, quality, access, and cost. These decisions have differential impacts on various groups. Regardless of who makes these decisions, however, water managers who want their decisions to be seen as legitimate must take into account the views, opinions, and outlooks of those affected by them. This has always been true. In ancient cities, water laws evolved to prioritize public, common uses first, followed by private needs (Boatright et al., 2004). Settlements often had conflicts with one another over water supply as increasing aridity took hold (for example, Nile basin settlements in the third millennium BC). The relevance of this equity principle remains important in modern cities – ranging from such diverse urban clusters as Los Angeles, New York, Mumbai and Mexico City, among others – where pressing decisions regarding water supply, drought response, and demand-management have become acute.

A major difference between ancient and modern cities as regards decision-making is that while sharing decision-making authority is unique to contemporary urban water decisions, in both ancient and contemporary times, decisions regarding the outcomes of these decisions are often based on equity criteria. Athens, Rome, and Barcelona again exemplify this. Athenian culture promoted water provision for baths and bathing, particularly in the context of gymnasia or schools for the education of principally male youth. While a public good associated with a communal expectation, homes of the affluent also had private facilities for bathing (Connolly and Dodge, 1998).

Rome, by contrast, was an innovator in viewing baths as well as lavatories as public goods that were the obligation of the state – and of aristocratic classes – to provide for the pleasure and edification of the general population. By the early fifth century AD, Rome provided some 11 imperial baths (thermae) that could accommodate large numbers of people of both genders at one time. While of great archeological, architectural, engineering, and aesthetic interest even today, from the standpoint of urban water policy, these baths – and their surrogates in other Roman cities and towns (for example, Barcelona) – reveal much about the notion of water equity in urbanized parts of the empire.

Aside from being fed by public aqueducts and in some instances special reservoirs, they were important social meeting places, and the considerable effort taken to make them attractive and ornamental, as well as utilitarian, symbolizes the considerable importance placed by authorities upon water equity – and on the priorities to be accorded to public water uses. From public fountains, water was supplied to public baths, and from there runoff went to the baths' own toilets, and then to public latrines distributed throughout the city (see Figure 2.1). Most important of all, only after these designated public uses were satisfied could individuals pay privately for uses diverted to their own homes (Connolly and Dodge, 1998; Coarelli, 2007; Boatright et al., 2004: 389–91).

While the case of Rome and Athens are better known than in other cities, the importance of these equity issues elsewhere has also been well documented in other ancient metropolises. For example, Constantinople, which in the late fourth century approached Rome in population, and later superseded it as Rome's population declined (Constantinople had upwards of half-a-million residents), was viewed by contemporaries as "enfeebled" by lack of water (Lenski, 2002: 395). Due to the concerted efforts of the emperors Valentinian and later, Valens, an ambitious program of civic improvements was undertaken to correct this deficiency. The latter constructed the first large aqueduct for the city during his reign. Both emperors were described as "obsessed" with the construction of waterworks, and concerned with projects of benefit to the general population (Lenski, 2002: 395).

Figure 2.1   Urban water management in Ancient Rome: public latrines (Ostia), baths (Pompeii)

In sum, three themes emerge with regard to equity and the ancient city. First, water was viewed as a public good important to commodious living. Second, the state was obliged to provide water services to as many residents as possible. And third, hygiene, health, and collective well-being were considered important goals of cities in ancient times that could be satisfied, in part, through the provision of water.

Infrastructure Investment and Long-Term Aspirations

As we have seen, the management of water supply and waste-related issues has been an important element in ensuring urban stability, development, and growth since antiquity. But how do growing cities affect outlying regions from whence their supplies are often acquired, and how do these outlying regions react? These issues continue to prompt environmental and societal challenges.

Two U.S. cities, Los Angeles and New York, are instructive cases. Early in their histories these cities, in their quest to acquire water, adopted a hegemonic relationship with their neighbors. In effect, they sought to control regional sources that could satisfy current as well as projected water needs (Hundley, 2001; New York City, 2011; Koeppel, 2000). Over time, both cities embraced collaboration with adjacent communities to address water supply and quality issues whose scope and impact required regional accommodation and sharing of authority.

New York's central challenges currently revolve around managing water quality and the safety of its drinking water. Meeting these challenges is virtually impossible without cooperation with non-governmental actors in other political jurisdictions from whence its water supply comes, and who would be severely burdened financially if the city had to build a large regional water filtration plant. For Los Angeles, by contrast, water (and air) quality issues in the Owens Valley – the source, since 1913, of one-third of the city's water – have driven efforts to partner with valley stakeholders to negotiate gradual reductions in flow and restoration of the watershed, efforts that were largely driven by court orders following law-suits against the city. While both cities were initially concerned with water supply, over time they both became increasingly worried over water quality and the need for integrated approaches to managing supply and quality.

Early in their histories, both cities faced many of the same challenges to public health and wastewater management that their Third World megacity counterparts face today. These included confronting the role that foul and unhealthy water plays in the spread of infectious disease (a particular problem for New York City which, in 1832, suffered a severe cholera epidemic attributed to contaminated drinking water: Koeppel,

2000; American Museum of Natural History, 2011). Moreover, in diverting water from outlying hinterlands, Los Angeles and New York generated well-documented, but vastly different, environmental and social impacts upon these adjacent regions. In the case of Los Angeles, diversion of water imposed reductions of both in-stream flow and groundwater in Owens Valley, degrading local fisheries and wildlife habitat (McQuilkin, 2011), while acquisition of adjacent lands overlying aquifers deprived Owens Valley communities of the ability to pursue real estate development for commercial and residential use (VanderBrug, 2009).

By acquiring much of the open space surrounding its reservoirs in the Catskill and Croton watersheds, a positive economic outcome generated by New York City was retention of low-density residential development that preserved the region's rural character (Westchester County Department of Planning, 2009). Later, sewage plant outfalls and non-point pollution around these same reservoirs released contaminants into the city's water supply which generated further, less popular land acquisition measures to avert pollution through eminent domain and condemnation suits, a strategy that continued through the early 1990s (New York State Department of Environmental Conservation, 2010a; 2010b).

From its founding in 1781, and for nearly a century afterwards, the Los Angeles River was the city's major water source. Prior to European (that is, Spanish) settlement, the Native American tribes in the region – collectively referred to as "Gabrielino" by the Spanish – lived in close harmony with the environment. Generally understanding the seasonality of rainfall and reliant for the most part on coastal resources, they had a "symbiotic" relationship with nature and used water communally. Native claims were forcibly ended shortly after conquest (*c.* 1770). The first families who founded and settled the "pueblo" almost immediately set about constructing a brush "toma" or dam across the river, diverting water into a so-called "Zanja Madre," or mother ditch, which fed homes and irrigation canals into fields that, at first, were closely adjacent to the plaza – the civic center of the early settlement (Los Angeles Department of Water and Power, 2010b).

In addition to developing an extensive system of *zanjas*, hand ditches which diverted water from streams and rivers to irrigate small farms and orchards, large estates, pueblos and missions, city officials sought ratification of a so-called "Pueblo" water right: an entitlement under traditional Spanish law to lay claim to all needed waters in the vicinity. Spain's King Carlos III granted control of the Los Angeles River to the infant city in 1781 (Los Angeles Department of Water and Power, 2010b). Various legal efforts were exercised in the nineteenth century – under Mexican and later U.S. rule – to ensure that this Pueblo water right was legally perfected. By the time the City of Los Angeles was incorporated in 1850, the city of

1600 was vested with all of the rights to the water of the river (Hundley, 2001).

In 1854 the *zanjas* became encapsulated into a city department that, in a few years, was leased to a private company and became the Los Angeles City Water Company. In 1902 the city purchased the company and begun work on an aqueduct. Privatization was largely a reaction to the high costs of maintaining and repairing the *zanjas* which, by the late nineteenth century, had grown to encompass ditches and channels, as well as water wheels to lift the water to gravity flow irrigation systems, along with some 300 miles of water mains, reservoirs, infiltration galleries and pumping facilities.

Early New York relied on domestic water supplies obtainable from sources in the immediate vicinity – from 1624 onward. Initially, these consisted of shallow, privately owned wells. Under Dutch rule, a period of about forty years, sanitation and water quality were decidedly poor: accumulations of human and animal waste were common, contaminated runoff into holding ponds was frequent, and there was no concerted effort to regulate harmful activities impinging on locally adjacent well-users (Koeppel, 2000). Thus, New York's initial efforts to develop a water supply system were largely animated by three concerns: accommodating population growth, averting communicable disease, and achieving both objectives while saving money (Glaeser, 2011: 99).

Under English rule in New York City (1664 and after), improvements were marginal. Foul, standing water was common, and outbreaks of epidemics stemming from poor water quality – including yellow fever and cholera – were not unknown (Koeppel, 2000). In 1677 the first general public-use well was dug near the fort at Bowling Green, while the first city reservoir was constructed on the east side of Broadway between Pearl and White Streets in 1776, about the time the city's population grew to over 20 000 residents. Initially, water pumped from wells near the Collect Pond, east of the reservoir, was distributed through hollow logs laid along main thoroughfares in Manhattan (New York City, 2011).

As the city grew, pollution of wells became a serious problem, as did periodic drought. These problems led to concerted efforts to supplement local supplies through cisterns and springs in upper Manhattan (a less developed area). Following the outbreak of a yellow fever epidemic in the last decade of the eighteenth century (1798) New York sought a safer, more secure, and disease-free water supply. In some respects this was a surprising early development, since widespread recognition of the relationship between contaminated water and communicable disease outbreaks were not really well understood until over half a century later – mostly through the work of John Snow (see, for example, Miasma Theory, n.d.).

During a cholera epidemic in London in 1848–49, Snow proposed the theory that the disease was caused by a particle that was ingested orally, rather than by a befouled component of miasmatic air. During a later epidemic (1853–54), he collected data supporting the view that cholera was primarily spread by sewage-contaminated water – mostly in South London.

At any rate, in 1800 the Manhattan Company (forerunner of Chase Manhattan Bank) sank a well at Reade and Centre Streets, pumped water into a reservoir on Chambers Street and distributed it through wooden mains to a portion of the community. This venture became, in effect, the city's first quasi-public water utility and was a major enhancement to the earlier Collect Pond (Willensky and White, 1988: 18). In 1830, in an effort to enhance emergency supplies, the city built a tank for fire protection at 13th and Broadway, which was replenished from a well. Water was distributed through 30.48 cm (12-inch) cast iron pipes.

As in Los Angeles, these efforts to provide a safe and secure supply were largely privately funded and managed (Koeppel, 2000). Led, ironically, by two New York statesmen who soon became mortal enemies – Aaron Burr and Alexander Hamilton – the city's Common Council was persuaded to obtain state legislative endorsement of the Manhattan Company's charter. Burr and Hamilton had different motives in advocating the charter and the company: the former sought financial profit through transforming the "surplus" revenues of the firm to his own design, while the latter was swayed by the desire to unburden city residents of a tax-supported public system (Glaeser, 2011: 99).

After weighing various alternatives for obtaining additional water supply, Los Angeles and New York expropriated distant sources. The contentious history of Los Angeles' efforts to acquire the Owens Valley, in contrast to those of New York in its Croton and Catskill watersheds, has been well documented (for example, Walton, 1993; Davis, 1993; Mulholland, 2002). New York officials sought to impound water from the Croton River, in today's Westchester County, and to build an aqueduct to carry water from what became known as the "Old" Croton Reservoir to the City. In contrast to Los Angeles, the urgency of an aqueduct was not as readily apparent to many local residents, and initial political support was far from unanimous. A major fire in 1835, which consumed a sizable portion of what is now lower Manhattan, convinced many wavering citizens of the need for an aqueduct (Koeppel, 2000). Moreover, even after the aqueduct was completed – in 1842 – not all city water users chose to connect themselves to the system, preferring to rely upon less reliable, but still cheaper, local supplies from wells and cisterns (Koeppel, 2000).

New York City's original aqueduct, known today as the Old Croton

Aqueduct, had an initial capacity of about 90 million gallons per day and was brought into service in 1842. Newer reservoirs were subsequently constructed to increase supply: Boyds Corner in 1873 and Middle Branch in 1878 (New York City, 2011). In 1883, as the city's continued growth and commercialization taxed this supply source, a commission was formed to build a *second* aqueduct from the Croton watershed together with additional storage reservoirs. This conduit, known as the New Croton Aqueduct, was brought into service in 1890, while still under construction. One of the biggest land use issues was the need to acquire land and right-of-way for the New Croton Dam and Aqueduct System – an effort begun in 1880 when 7000 acres were acquired to harness the Croton River's three branches, while a 20 square mile area was needed by the city on which to build the New Croton Dam. Twenty-one dwellings and barns, one and a half dozen stores, churches, schools, grist mills, flour mills, saw mills, four towns, and over four hundred farms were condemned and taken over to build the dam – and some 1500 bodies were removed from six cemeteries and relocated along with their stones and fences. One local historical account states that "protests, lawsuits and some confusion preceded payment of claims" (Village of Croton, 2010).

At the same time, the municipal system was consolidated from various water systems in the communities now consisting of the Boroughs of Manhattan, the Bronx, Brooklyn, Queens and Staten Island. An important parallel with Los Angeles is how water system consolidation became an important first step toward *municipal* annexation. For Los Angeles, completion of the first Owens Valley Aqueduct in 1913 leveraged the city's ability to force smaller communities coveting water (for example, Hollywood) to accede to annexation as a condition for becoming connected to the distribution system.

A third phase development occurred in the early twentieth century. In 1905, a Board of Water Supply established by the New York State Legislature cooperated with the city in developing the Catskill region as an additional water source – with the former planning and constructing facilities to impound Esopus Creek, and to deliver the water to the city via the Ashokan Reservoir and Catskill Aqueduct – a project completed in 1915 and given over to the City's Department of Water Supply, Gas and Electricity for operation and maintenance. The remaining development of the Catskill System, involving the construction of the Schoharie Reservoir and Shandaken Tunnel, was completed in 1928.

A fourth and final action to acquire water was the effort to allocate the Delaware River. In 1927 the Board of Water Supply submitted a plan to the state Board of Estimate and Apportionment for the development of the upper portion of the Rondout watershed and tributaries of the

Delaware within New York State. This project was approved in 1928. Work was subsequently delayed by an action brought by the State of New Jersey in the U.S. Supreme Court to enjoin the City and State of New York from using the waters of any Delaware River tributary (New York City, 2011). This case underscores the regional animosity brought about by the City's effort to seek water hegemony.

In 1931 the Supreme Court upheld the City's right to augment its water supply from the Delaware's headwaters. However, a second Supreme Court ruling, in 1954, was required to adjudicate riparian allocation of the Delaware between New York, New Jersey, and Pennsylvania (Derthick, 1974: 48, 54). Construction of the Delaware System was begun in March 1937 and entered service in stages: the Delaware Aqueduct was completed in 1944, Rondout Reservoir in 1950, Neversink Reservoir in 1954, Pepacton Reservoir in 1955 and Cannonsville Reservoir in 1964. Figure 2.2 depicts the current New York water supply system.

Los Angeles followed a similar path in its efforts to build a major supply conduit from the Owens Valley. As the city's population rapidly grew after 1880, it became apparent that the Los Angeles River was simply not large enough to support the city's transformation into a large metropolis. Its population doubled during the 1890s, from 50 000 to 100 000, and more than doubled again within five years (to over 250 000), all but depleting local groundwater. Moreover, the city's incorporated area doubled between 1890 and 1900 as nearby cities embraced annexation to ensure water.

Fred Eaton, one-time city engineer during the 1890s, Mayor from 1899 to 1901, and superintendent of Los Angeles' municipal water system conceived of an Owens River aqueduct in the early 1900s (Davis, 1993: 5–9). After an unusually harsh drought in 1904, William Mulholland – a protégé of Eaton and now city engineer – asked his mentor to "show me this water supply" in the Owens Valley about which Eaton had often spoke. Following an intrepid journey both took through the region, which included a preliminary survey of an aqueduct route, events moved quickly. In September 1905, voters approved by a 10–1 margin a $1.5 million project to acquire right-of-way, and to build an aqueduct that would stretch from north of Independence some 376 km (234 miles) southeast to the San Fernando Valley, a recently incorporated area of the city.

At precisely the moment political forces in Los Angeles maneuvered to acquire Owens Valley water rights, the newly formed U.S. Reclamation Service drafted a plan to irrigate the Owens Valley by constructing one or more dams in the vicinity of Long Valley. As a federal agency man-dated to promote irrigation, the Service was inclined to support the people of the valley against those of a large city seeking to augment its

Source: New York City Department of Environmental Protection (2010).

Figure 2.2 New York City's water supply system

water supply. However, the Reclamation Service's southwestern regional chief, Joseph P. Lippincott served (secretly) as a paid consultant to Los Angeles, abetting the city's plans, since Lippincott advocated for the city's interests in Washington, DC, not those of the Owens Valley. Lippincott also helped ensure that, while valley lands would be set aside for public purpose, no land rights would be secured: an action that abetted Eaton's efforts to set about buying up options on land for aqueduct construction (Kahrl, 1982).

Within two years, two other efforts were completed that worked in the city's favor: a successful campaign to obtain Congressional approval of the City's application to build the aqueduct was effectuated in June 1906; while in 1907, Los Angeles voters approved a second bond measure authorizing $23 million for aqueduct construction. Construction began in 1908 and the project was completed in November 1913.

Like New York City, the Owens Valley was one phase in the city's water supply expansion. By the early 1920s, the Board of Public Service commissioners (the overseers of the Los Angeles Department of Water and Power or LADWP), became aware that the city would exceed the Owens Valley's supply by 1940 (thus, a second aqueduct was built in the Owens Valley all the way to Mono Lake, a project approved by voters in 1930 and completed in 1940).

A third phase began with the efforts of Mulholland to acquire water from the Colorado River. A four-year series of surveys began in 1923 to find an alignment that would bring the water of the Colorado River to Los Angeles. In 1925 the Department of Water and Power (LADWP) was established, and the voters of Los Angeles approved a $2 million bond issue to perform the engineering for the Colorado River Aqueduct. While the six cooperating states of the basin sought a means to allocate the Colorado's flow, an effort that began with the 1922 Colorado River Compact and required Congressional passage of the Boulder Canyon Dam Act in 1928, Los Angeles also acted on its own.

In 1928 the city and LADWP got the state legislature to create the Metropolitan Water District of Southern California or MWD (Fogelson, 1993: 101–103; Erie, 2006). In 1931, voters approved a $220 million bond issue for construction, and work began on the ten-year, 300-mile long project which, in conjunction with the State Water Project that is also managed by the Metropolitan Water District, now supplies about 48 percent of Los Angeles, Orange, Ventura, San Bernardino, Riverside, and San Diego Counties' water. In the 1970s the regional cooperative also began importing water from Northern California via the State Water Project and the California Aqueduct. Figure 2.3 depicts Los Angeles' water system.

Figure 2.3   City of Los Angeles' water supply system

Subsequent to completion of their respective aqueduct systems, both cities began to face a series of water-related environmental quality challenges that required unprecedented levels of regional collaboration to resolve. In Los Angeles' case, this collaboration emerged after a series of litigious actions resulting from adverse ecological and tribal-equity issues. In New York, they came about through harsh economic realities brought to the fore by a severe federal regulatory challenge.

As far back as 1913, the draining of Owens Lake as a result of the opening of the first Los Angeles Aqueduct exposed the alkali lake bed to winds that lofted toxic dust clouds containing selenium, cadmium, arsenic and other elements throughout the region. Airborne particulates were often suspended for days during excessively dry periods, and have long posed a health hazard to local residents. They have even posed risks to communities further to the south. In the 1970s, the siphoning off of additional flows following completion of a second and larger aqueduct ignited further protest.

These environmental impacts to Owens Lake, Mono Lake, and other watersheds in the Owens Basin (for example, Lee Vining, Walker, and

Parker Creeks) dovetailed with concerns regarding water management in Los Angeles itself, beginning in the 1970s.

In 2013 a historic "Stream Restoration Agreement" was concluded between Owens Basin officials and the city of Los Angeles. The agreement required the city to restore stream flows to historic levels. The agreement was long in coming about, driven largely by a series of court decisions brought about by protracted litigation, changes to state regulations, and lengthy negotiations, all of which were pivotal in bringing about this major policy change. While partly aided by the authority of science, voter petitions, and fund-raising drives, vocal public protest was also a key (Mono Lake Newsletter, 2014; Walton, 1993).

Continuing drought and unrelenting population growth compelled the city to embrace a more adaptive approach to water management reliant on conservation, drought management, and a balance between augmenting supplies while providing incentives to lower demand: a method termed *integrated water resource management*. This approach relies on incentive-based methods to reduce water use and has been facilitated in part by concerns over climate change as well as the stresses and strains felt throughout its water importing regions (Los Angeles Department of Water and Power, 2010a).

For New York, one of the benefits of the city's acquisition of the Catskill and Croton watersheds during the nineteenth century was the opportunity to ensure a virtually pristine source-water strategy. The city's reservoirs are surrounded by hardwood and evergreen forests that naturally filter water, retard erosion, and avert sedimentation that would otherwise reduce drinking water quality. They also marked New York's distinction as the nation's largest city *without* a drinking water treatment plant until relatively recently (American Planning Association, 2011).

CODA: THINKING ABOUT AN URBAN ECOLOGY OF WATER

As we prepare to enter the third decade of the twenty-first century, cities worldwide face the challenge of trying to furnish their citizens with a resilient system for water management in the face of severe climate variability, including drought – and to do so while simultaneously facing significant challenges to the building of additional infrastructure. The need to understand urban history in this endeavor remains critical (Box 2.1) because past precedents, traditions, and patterns of behavior often shape and determine the manner in which later choices are made. As we have seen, this history

---

BOX 2.1   WATER HISTORIES AND PATH DEPENDENCY

A Research Agenda for Understanding Urban Water History

- How does time and history (path dependency, policy feedback, timing and sequence, punctuated and incremental change) affect outcomes? Sub-optimal local institutions and governance arrangements may become embedded in regions, with outcomes highly sensitive to initial conditions.
- 'Recent government urban policy has been developed with seemingly little understanding of the origins of urban planning or why Australian cities take their present form and structure' (Pat Troy).
- What do we need to know about this history? Are changes in trajectory plausible?

Source: Courtesy of Professor Lionel Frost, Monash University, Melbourne, Australia.

---

is comprised of infrastructure, laws and rules, and institutions. As we shall see in Chapter 3, these three things continue to comprise the major elements of water management in cities, and constitute key elements in their water futures.

3. Roles for civil engineering, law and institutions in urban water management

Water resources development has been an important vehicle for economic expansion in cities throughout history and a driver of population growth (Willoughby, 1999). In the U.S., cities like Los Angeles annexed neighboring communities to stage ocean-going commerce and persuaded the federal government to support harbor construction, while San Francisco, Denver, and Atlanta sought federal support for their efforts to secure their position as multi-modal transportation hubs and major regional economic hegemons (Pomeroy, 1965; Starr, 1985; Hundley, 2009; Logan and Muse, 1989). This pattern has also been true in Australia, where urban form shaped, and was shaped by, the freshwater environment in places such as Adelaide, Melbourne, and Sydney (Frost, 2013).

Cities have a water history – but they also have a water anatomy. Like living creatures with circulatory systems, every city has developed its own unique "plumbing" system with water supply and wastewater treatment infrastructure, and elaborate systems of law and regulatory institutions to govern and manage these systems. This chapter examines this anatomy and physiology by focusing on some urban examples that typify the range of both conventional and novel approaches to urban water management in our time. Our objective is to trace various ways in which cities seek to attain water resilience. The principal lessons drawn come directly from our PIRE project and our investigations into measures adopted by Melbourne, Australia.

Before defining resilience, we need to consider another concept, often used unreflectively as a synonym: adaptiveness. We define resilience as the capacity of an urban area (or any water user system) to undergo shocks from climate change or other environmental stressors while retaining essentially the same function, structure, feedbacks and identity as before (see Kiparsky et al., 2013; Brown et al., 2009; Dobbie et al., 2014). By contrast, adaptiveness, or the ability to experience "adaptive transitions" as some writers phrase it, is the capacity to respond to change in sometimes unplanned and improvisational ways.

While resilience implies a top-down and centralized institutional approach to coping with pressures, adaptiveness suggests the possibility of a "bottom-up" and decentralized capacity to evolve, rather nimbly, from one condition or state to another. In actuality, few cities have experienced adaptive transitions (Ferguson et al., 2013; Kiparsky et al., 2013; Gleick, 2003). This is because achieving this objective requires rethinking not only the design of engineered infrastructure but law and administrative institutions as well.

A TALE OF TWO MEGACITIES AND A THIRD GROWING METROPOLIS

In recent years, several large cities in developed and developing countries alike have instituted measures to secure more adaptive – as opposed to simply resilient – water systems by moving away from traditional modes of top-down operations and decision-making and toward more bottom-up approaches. While partly driven by climate concerns, the immediate drivers of these adaptation efforts have been population growth, the need to share supplies with neighboring communities, and demands to restore threatened or endangered habitat. In most cases, these efforts have been spurred by some type of crisis – such as rapid population growth, drought, and conflicts with neighboring communities and outlying regions. While we could cite many examples, three large cities – representing an array of baseline climates – typify some of these drivers: Tokyo, Mexico City, and Melbourne.

Tokyo is located in a traditionally wet climate, receiving some 60 inches of precipitation per year, most of which falling as rain during mid-summer. Under normal circumstances, we might assume with such an abundant, natural supply of precipitation, the city has little to worry about as regards a stable and secure water supply – even with a population now exceeding 13 million. However, as with most major cities, Tokyo has faced many challenges affecting both water supply and urban demand. Population growth, changes in life style, economic growth (which has become practically incessant since the end of World War II), and uncertainties affecting supply are having significant impacts on the city. In general, despite a growing appreciation for the long-term effects of climate change on the city's water supply, demands for water continue to increase (Tokyo Waterworks Bureau, 2013).

Practically from the moment Japan sought to modernize its society, in part through mimicking the technologies of the West, it also sought to modernize and develop its provision of urban water supplies. During the early period of the Meiji restoration (1898), the city of Tokyo's waterworks began supplying water from a modern treatment facility, the Yodobashi purification plant. Ironically, investments in sewage infrastructure came about much later, a fact which explains high disease rates, especially for cholera, in comparison to many European cities, for example (Otaki et al., 2007).

Today, Tokyo has one of the world's most sophisticated and "smart" metered and regulated water supply systems. Tokyo's waterworks are among the world's largest, and are noted for reliance upon the highest levels of technology (see Figure 3.1).

After World War II, rapid in-migration and economic growth dramatically increased water demands at the same time as planners decided to pave over small waterways to facilitate urban expansion. Increased consumption led to declines in groundwater and land subsidence. Since the 1980s, climate change concerns, including local "heat island" effects from urbanization leading to additional energy use, have prompted introduction of large-scale wastewater reuse, non-potable stormwater harvesting, groundwater withdrawal restrictions, and aggressive conservation. As a result of the city's efforts to foster conservation, it has been estimated that per capita household demand will plateau within the next decade.

Consistent with approaches to technological innovation in other domains, Tokyo's Metropolitan Waterworks Bureau has sought to develop new projects through active international cooperation, and from an international perspective – both to help institute globally recognized standards for engineering, and to model approaches that can be exported to less developed nations where needs for clean water in urban areas especially are crucial (Tokyo Waterworks Bureau, 2013).

In line with developments across the globe, Tokyo's Bureau of Waterworks has been strongly advocating – and implementing – a variety of innovations to conserve as well as protect potable water supplies. Highly advanced water treatment systems have been installed in virtually all the city's potable water treatment plants, particularly in the large Tonegawa River system, and an extensive public education and outreach program, Tokyo Tap Water's "Water for Life," has been introduced to help customers better understand their role in conserving water and protecting its quality, and to comprehend how waterworks services operate. Meanwhile, systematic replacement of components of the city's huge water distribution system have been taking place on a regular and sustained basis – to a much higher degree than is experienced in, say, U.S. cities. Finally, considerable efforts have been taken to publicize the overall public health, ecological, energy savings, and cost savings benefits associated with tap water use as opposed to bottled water, issues taken very seriously in Japan as a whole.

Mexico City, one of world's largest cities (> 20 million), and recipient of

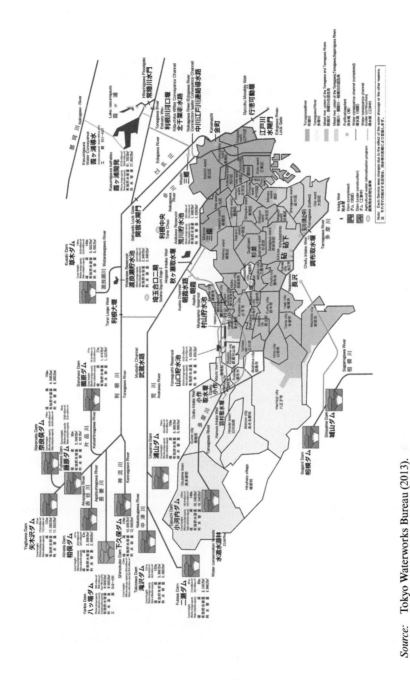

Source: Tokyo Waterworks Bureau (2013).

Figure 3.1 Tokyo's water distribution system

some 28 inches of rainfall annually, has instituted some of the most ambitious plans to harvest water supplies from outside the metropolitan region. Most of the Federal district receives its water supplies from the Cutzamala system, one of the world's largest water supply systems in terms of both the total quantity of water supplied (about 485 million cubic meters/year) and of the 1100 meters (3600 feet) difference in elevation between source and users. Begun in the 1970s, the $1.3 billion supply system (a price tag higher than national investment in Mexico's entire public sector at the time; Tortajada and Casteian, 2003) was built in stages, in some instances incorporating earlier projects dating to the 1940s. Delivery to the Valley of Mexico from more than 150 km away is achieved via a system of seven dams and storage reservoirs, six pumping stations, open channels, tunnels, pipelines and aqueducts – and a water purification plant (Figure 3.2).

The entire system supplies almost 20 percent of the Valley of Mexico's total water supply (some 82 m³/s). And, it is not only large, but highly energy intensive, consuming some 1.3 and 1.8 terawatt hours a year, equivalent to 0.6 percent of Mexico's total energy consumption, and representing a cost of about $65 million a year – necessary in order to pump water some 1100 meters from the lowest to the system's highest point (from where gravity flow takes over). Many communities had to be relocated to make way for this massive infrastructure, and unresolved disputes over proper compensation for losing homes continue to simmer (IMTA, 1987).

Despite the ambitious scale and scope of this project, Mexico City continues to face some of the most severe challenges in potable water provision of any major city in the world. Climate change is very much a motivator for further action, while continued population growth pressure is another. Other options being pursued to address these challenges include efforts to reduce residential water demand, using more reclaimed wastewater for local agriculture and non-potable uses, and employing stormwater capture for groundwater recharge and some community uses. Given the availability of public investment funds and low water tariffs charged in Mexico, the likely effectiveness of these measures is subject to considerable debate.

A National Water Commission (Conagua) convened some six years ago concluded that 40 percent of potable water nationwide was being lost through leaks in urban provision systems, while another 20 percent was "unaccounted water loss" through billing errors and illegal water connections – problems said to be characteristic of Mexico City and other metro areas (Geo-Mexico, 2013). While the commission has recommended a number of reforms, including a new metropolitan decision-making body supposedly empowered to choose "which sources of water will be used, set timelines and commitments, and monitor all activities carried out under the plan," a continuing challenge in implementation remains artificially

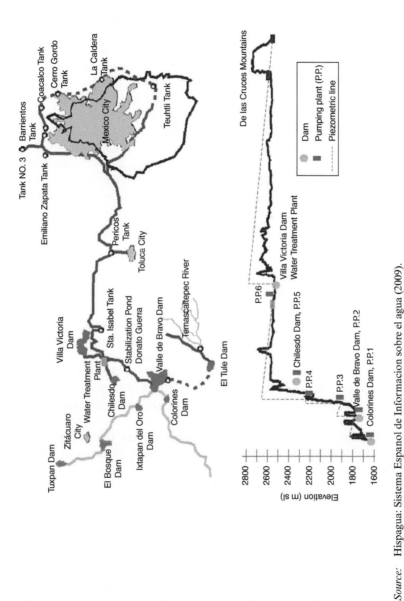

Source: Hispagua: Sistema Espanol de Informacion sobre el agua (2009).

Figure 3.2 Mapping the water supply infrastructure of Mexico City

low water prices, which historically have failed to account for the actual costs of water production, maintenance, and service delivery.

It is stated that consumer water rates charged in the Valley of Mexico cover only half of the true costs of service provision (Geo-Mexico, 2013). Partly as a consequence of low rates, maintaining the Cutzamala system has been an ongoing challenge. Maintenance and repairs tend to be performed over holidays when demand is low. Since 1993, a parallel network of canals and pipelines has been built alongside the original system, allowing for sections of the system to be shut down for maintenance, and obviating the need to close it down when work occurs.

Related to these issues is a larger set of national constraints that affect not just urban water policy, but Mexico as a whole. The nation suffers from severe water vulnerability based on both high urban and agricultural demands on one hand, and growing urban uses for which infrastructure and supply availability falls short (Padowski et al., 2015). In addition, under-reporting, as well as "illegal water use, poor surveillance and lack of enforcement" (Medellin-Azuara et al., 2013: 21) are among several institutional deficiencies in water resources management.

The Unique Case of Melbourne

Melbourne, Australia is home to some four and a half million people, practices some of the most innovative water management approaches of any large city, and experienced a severe challenge to water supply in the late twentieth and early twenty-first centuries. The Millennium Drought (1997–2009) afforded an opportunity for implementing policy and infrastructure innovations that may have worldwide application, as we will see later on (Ferguson et al., 2013; Low et al., 2015). The worst drought since European settlement began in 1788, the Millennium Drought's effects were felt across ecosystems, agriculture, the economy and society. Before the drought ended, reservoir storage volumes fell to a historic low of 25.5 percent capacity by June 2009. And yet, over the drought's 12-year period, Melbourne reduced its per capita water demand by almost 50 percent. Key to understanding Melbourne's adoption of technical innovations – as we shall see – are institutional changes in the water sector at national, state and local levels that have facilitated their adoption (see Chapters 4 and 8).

A Focus on Engineering: From Conventional to Novel

Melbourne's water supply is a complex interconnected system of 10 storage reservoirs, with a total capacity of 1812 GL, over 40 service reservoirs,

160000 hectares of catchments, and a transfer system comprising hundreds of kilometers of pipelines, tunnels and aqueducts.[1] Most of the water catchments are forested and closed to the public, including the two largest catchments, the Yarra and the Thomson. The largest reservoir, the Thomson Reservoir, was completed in 1984 and was intended to make the city "drought-proof" after the 1982–83 "Short but Sharp" Drought.

Catchment water is traditionally an inexpensive source of water for the city due to the minimal need for transfer pumping or treatment but, being climate dependent, can be highly variable in quantity. Relatively little groundwater is used in Melbourne. Groundwater licenses are capped at 33 GL per year, and most is used for irrigating market gardens and golf courses. Presently, Melbourne's annual water use is around 410 GL of potable water, 21 GL of recycled water, and an estimated 10 GL of stormwater and rainwater harvesting. Melbourne's supply system consists of over 24 000 km of water mains, with a level of leakage (non-revenue water) that is low by international standards (9 percent in a typical year).

LAW AND INSTITUTIONS IN THE WATER-SUSTAINABLE CITY: SOME LESSONS FROM PIRE

The ability to adopt technical and other innovations in response to the Millennium Drought was facilitated by institutional and legal features unique to Melbourne's setting and circumstances. At the same time, however, Melbourne is a fascinating case precisely because the city and its surrounding region maintained elements of traditional water supply approaches (for example, a large diversion project, a desalination plant) as well as more adaptive approaches to the Millennium Drought. As one local water official noted in retrospect: economic, social, and environmental outcomes were collectively considered together, local agencies were under close scrutiny regarding their decisions, there was wide acceptance that the conditions being managed were the "new normal" for the foreseeable future, and that improvements needed to be made on several fronts – to water infrastructure, user efficiency, and augmentation of supply (Doolan, 2015).

Melbourne's water companies are constituted under the Victorian Water Act of 1989. State Ministers of Water, Environment, Health, and Treasury collectively oversee the water sector. A Department of Environment and Primary Industries (DEPI) supports the Minister for Water and the Minister for the Environment, while an Essential Services Commission regulates the water sector. Retailers must submit plans to the Commission to justify rate increases, while dividends are paid annually to the State Treasury. A clause in the Water Act allows the Minister for Water to issue a Statement of

Obligations (SoO) in relation to water companies' performance and it also requires them to adopt a joint Drought Response Plan (DRP).

While the DRP specifies various levels of water restriction based on water storage levels, the Minister for Water makes or lifts restrictions with the advice and input of water companies' directors. The latter, as well as DEPI, also influence other drought responses, reflecting Victoria's deep interest in projecting strong leadership, while the Council of Australian Governments' National Water Reform Framework of 1994 promotes a nationally integrated approach to water management. In short, this framework helped assure that when the Millennium Drought began in 1997 Melbourne could readily and quickly introduce supply and demand-side measures.

PUBLIC VERSUS PRIVATE CONTROL IN CITIES

Private ownership of freshwater provision and treatment is one of the world's most rapidly growing businesses, and is becoming especially popular in large cities. While many reasons for privatization have arisen, among the most important is the need for capital to meet growing demands for clean water in developing nations. These demands, and the role of privatization in meeting them, have power, equity, infrastructural, and aspirational implications. While private versus public control – despite widespread interest among water scholars – is an important global issue, it is, somewhat surprisingly, not a critical issue on the cases discussed in this chapter. What might that tell us about the private–public debate?

First, public and privately owned providers have several similarities in common. They both exercise a monopoly over water provision in their regions of operation. The greater the area covered, and the more exclusive the rights to cover it, the cheaper it is for a water provider to build, maintain, and manage the infrastructure that serves that region. Second, the distinction between the two provision systems is sometimes overstated. Providers receive their supplies from state-owned reservoirs and groundwater basins, and water is delivered wholesale via dams, reservoirs, aqueducts, wells, and pumping stations built and maintained by government agencies. Moreover, the state usually administers these projects through its control of various river basin and groundwater-management authorities. In most cities, therefore, privatization is actually linked to some form of public monopoly-granting authority.

Third, the dramatic growth of privatization has prompted questions regarding political accountability, economic fairness, and the willingness of vendors to vigilantly prioritize public health and other community concerns above handsome returns on investment. Critics charge that public participation in decisions regarding pricing, service, and access is sharply limited because decisions are made by corporate boards rather than publicly accountable officials. Finally, private providers' operations are less transparent to the general public than is the case with public providers. Once control over local supplies and their provision are turned over to private companies, or even when plans to do so are contemplated, enterprises may shield their operations to protect proprietary interests. When this occurs, local political support for privatization often erodes among public officials. This situation has arisen in recent years in Pune, India; Karachi, Pakistan; Cochabamba, Bolivia; Manila, Philippines, and elsewhere. Transparency is often problematical. When control over urban water supplies and their provision is relinquished to private companies, it is often the case that these companies shield their operations from public scrutiny because they consider their products and services to be proprietary interests that must be protected from potential competitors. The well-documented and widely reported experience of Bolivia with privatization in the past two decades is a good example of this criticism. It is also a case that looms large in the annals of economic globalization of water rights.

Bolivia's privatization effort began in the early 1990s. A bi-national consortium of French and U.S. companies effectively took over control of supplies in several cities, including La Paz. Charges were imposed upon water taken from private wells, under the pretext that private vendors were given, by formal agreement and in the cause of greater efficiency, total control over local supplies. After protests erupted, one of the consortium's participants – Bechtel – abandoned its Bolivian water operations. Following similar unrest in La Paz toward the operations of a French-owned company, Aguas de Illimani S.A. (a subsidiary of Suez) also abandoned its Bolivian operations (Baer, 2008).

In Tokyo, Mexico City, and Melbourne many of these issues have arisen and the manner in which they have been managed is instructive. All three cities operate publicly owned water supply systems. The Tokyo Waterworks are operated as a local public enterprise, which is required to maintain economic efficiency. The water company characterizes itself as a waterworks and uses the enterprise's accounting "self-supporting accounting system," and the entire system's expenses are covered by revenue acquired by water charges (that is, tariffs). However, consistent with what we characterized as the "mixed" nature of many water utilities, capital costs for upgrading the water system are paid for by a combination of enterprise loans, governmental subsidies, and a portion of revenues transferred from the utility's general account. Finally, the Tokyo Metropolitan Assembly has the authority to make decisions on the budget, approve account settlement, and make revisions to low water charges. In short, while a "public" entity, Tokyo's water supply provider operates very much according to rules and principles characteristic of a privately owned business.

In the case of Mexico City, the tradition of public ownership coupled with low water rates is commonly attributed to be legacies of the 1910 revolution. In effect, water is considered a human right (a 2011 amendment to Mexico's constitution enshrines potable water as a basic human right) and tariffs are kept artificially low, as we have seen, and are heavily subsidized. While it would be easy to attribute high rates of urban water losses on such lackadaisical financial accounting practices, it also would be shortsighted. Despite low water rates, water "theft" is relatively common and unpaid water bills are infrequently punished. Moreover, it would be unfair to claim that Mexico City – and other urban communities – fails to appreciate the need for business savvy models of management. Water leaks are mostly caused by the age of the distribution system. Moreover, the recently created metropolitan decision-making entity for the Valley of Mexico is designed, in part, to alleviate any "possibility of misinterpretation in collaboration" (Geo-Mexico, 2013).

Finally, reforms are being introduced to waive water charges in areas receiving poor or sporadic service at last, while raising rates elsewhere to compensate, and to encourage reductions in consumption whenever possible.

Finally, Melbourne has long practiced water tariff systems that charge more for higher rates of water use. An open, democratic society with a long tradition of political activism and collaboration with various publics (an issue we will revisit in detail in the next chapter), Australians generally, and Melburnians in particular probably would not take to private water utility ownership. In short, as an issue related to control of urban water, the question of private versus public ownership has numerous implications for the advent of a water-sensitive city.

SUMMARY

The basis for the modern city is the control of water, and the management of water supply, in all its ramifications. This includes, at the most basic level, infrastructure to ensure that sanitary and public concerns over the health and safety of water are mitigated; that water supply can be provided reliably and affordably; and that hazards from too much water (that is, flooding) can be alleviated. As regards the latter, throughout history cities have faced the twin challenges of too much, or too little, water at inopportune times. This chapter elucidated the interconnections between

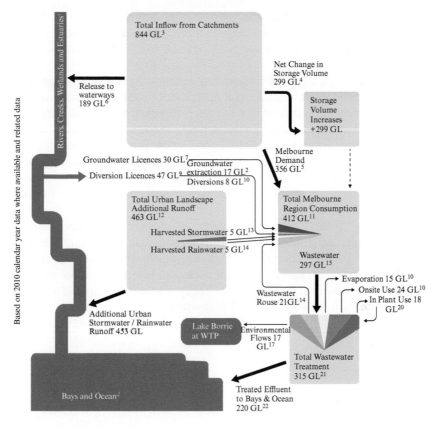

Note: See original figure for notes.

Source: Ministerial Advisory Council for the Living Melbourne, Living Victoria Plan for Water (2011).

Figure 3.3   Schematic of water inflows and outflows for Melbourne

the growth of civil engineering, law, and institutions in the management of urban water supply and treatment systems.

In considering the lessons of these various cities for understanding the urban ecology of water, we need to remind ourselves that every drop of freshwater has an important – really, indispensable – role in the life of cities. As conceptualized in Figure 3.3, which depicts Melbourne's water inflows from all sources, as well as water allocations, one can envision a city – most any city – as a kind of hydraulic "organism" whose components use energy to store, transport, treat, and dispose of water wastes – and which receives inflow from numerous sources. The latter is stored in surface impoundments as well as in recharged aquifers, consumes or uses a significant amount of it, and is finally released back into bays, estuaries, rivers, and finally – the ocean.

Eventually, this cycle of inflow and outflow will begin again as a continuous process. A fundamental difference between modern cities and those of antiquity, discussed in Chapter 2, however, is that the science of watersheds has advanced to the point where the volumes of water can be carefully measured and monitored – permitting, if we so choose, prudent and farsighted management. Another difference is that in thinking about water management in cities of the future, such a schematic permits consideration of a range of management options and alternatives for each management component.

Beginning with Chapter 4, which considers differences between contemporary and traditional alternatives to urban water management, we will show how consideration of these various options evolves over time, is accompanied by changes in institutions, and depends upon available innovations. We consider two cities already discussed in greater detail in this examination: Los Angeles and Melbourne.

NOTE

1. For a review of the data and infrastructure information discussed here see Low et al. (2015).

4. Divergent approaches: a typology of traditional and contemporary alternatives as seen in Los Angeles and Melbourne

The aim of this chapter is: (1) to review historical transitions in water management with a specific focus on two cities central to our PIRE project (greater Melbourne and Los Angeles); (2) to provide an analysis of current and emerging water policy innovations and management approaches they employ; and (3) draw generalizable lessons from these experiences. Our goal is to illuminate lessons for institutional and policy reform and technological innovation that can enhance water security in other cities.

Melbourne and Los Angeles are similar in some important water-related respects. They are modern, industrial-era cities that share moderate to semi-arid climates marked by periods of seasonal, episodic precipitation and prolonged, extreme drought. Each city also has a highly developed system of water management that illustrates the challenges in transitioning from conventional means of providing and treating water to a more integrated approach to its management. We begin by advancing a framework for understanding transitions, innovations, and policy reforms in urban water management. We then describe the historical development of major water management paradigms and water policies in each respective location. Finally, we examine current and emerging policies and management paradigms and provide illustrations of each.

A FRAMEWORK FOR UNDERSTANDING TRADITIONAL AND CONTEMPORARY APPROACHES

In general, major transitions between water management approaches have occurred in Western, industrialized societies during similar historical epochs. This is due to shared legal and cultural foundations, interdependent economies, and the affluence of Western societies. The latter directly influences a city's ability to acquire new technologies and management approaches to achieve higher standards of living. Several academics have characterized these urban transitions in their respective world regions (for example, Hundley, 2001; Van der Brugge and Rotmans, 2007; Crase, 2009; Hanak et al., 2012a).

In Australia, Rebekah Brown and her colleagues (2009) have advanced a popular and widely cited approach in which distinct periods, or archetypical "cities," can be depicted. Applied initially to water management transitions in Australia, the framework is applicable to most Western cities, we believe, including those in Europe and the U.S. This framework is based on the notion of a hydro-social contract – a concept used to describe the arrangements negotiated between government, industry, and society, and "shaped by the dominant cultural perspective and historically embedded urban water values, expressed through institutional arrangements and regulatory frameworks, and physically represented through water systems infrastructure" (Koo-Oshima and Narain, 2012).

Three pillars undergird this transitions framework and support the foundations of water management in any city. These are, respectively, regulatory, normative and cultural-cognitive pillars. These pillars are interdependent and mutually reinforcing – hindering, or facilitating, transformation from one epoch to another.

The regulatory pillar refers to rules, laws, policies, and regulations legally sanctioned by a recognized source of power capable of enforcing conformity, either through reward or punishment. It also includes less formal, unwritten mechanisms to influence behavior that induces guilt or fear, such as public shaming or ostracism. However, all regulatory processes, formal and informal, rely on the powerful force of coercion to achieve the desired effect of conformity.

The normative pillar refers to the norms and values present in society. Norms are social prescriptions specifying standards of behavior, while values are the "conceptions of the preferred or the desirable, together with the construction of standards, to which existing structures or behaviors can be compared and assessed" (Scott, 2013: 64). The normative pillar consists of formally and informally constructed social roles, social expectations and responsibilities, and appropriate practices (Scott, 2013). Finally, the cultural-cognitive pillar refers to the "shared conceptions that constitute the nature of social reality and create the frames through which meaning is made" (Scott, 2013: 67). This includes: how information is received by participants in the process of decision-making, and how it might be retained, retrieved, organized into memory, and interpreted.

PILLARS AND TRANSITIONS: CITIES AND GOVERNANCE

Pillars may combine and work to reinforce an institution, or be misaligned and work at odds. Institutions are characterized by their relative resistance to change (Jepperson, 1991) and are credited with providing long-term stability (Giddens, 1984). Moreover, institutions may be created, reproduced, and reinforced over time. However, they are also subject to gradual and sometimes radical change as their sustaining resources or the human interactions on which they depend are altered. Thus, institutions are not static and a transitional state exists during the process of institutional change from one water "regime" to another – somewhat along the lines of what we saw in the case of Melbourne: a fusion of older approaches and newer innovations.

The relationship between transitions and urban water management is as follows: as persistent social problems arise with respect to the management of water – its quality as well as quantity – "structural transformations" are necessary. These transitions are more transformational than the shifts that take place earlier in the evolutionary process, because they entail fundamental changes in water systems and their components. As seen in the cities of antiquity discussed in Chapter 2, for instance – as well as the contemporary examples discussed in Chapter 3, a transition can, in general terms, be portrayed as a long-term process of change during which a society or a subsystem of society fundamentally moves from one condition or state to another through the adoption of various innovations (Rotmans et al., 2001).

Transitions from one epoch to another can be guided or unguided. Unguided transitions are more organic and evolutionary in nature, in that "the outcome [of a transition] is not planned in a significant way" (Loorbach and Rotmans, 2006). By contrast, guided transitions (also referred to as goal-oriented, teleological transitions) have "(diffuse) goals or visions of the end state [that] are guiding public actors and orienting the strategic decisions of private actors" (Loorbach and Rotmans, 2006).

Moreover, transitions generally evolve through four phases (Rotmans et al., 2001: 17): (1) a *predevelopment phase* with little change in the status quo; (2) a *take-off phase* where the process of change gets under way as systems begin to shift; (3) a *breakthrough phase* where visible structural changes take place through accumulation of socio-cultural, economic, ecological and institutional changes that react to each other (in this phase, collective learning processes, diffusion (of innovations), and embedding processes occur); and (4) a *stabilization phase* where the speed of social change decreases and a new dynamic equilibrium is reached.

A major point of contention among academics who study these transitions is the issue of their management (Van der Brugge and Rotmans, 2007): who dominates a transition, and is its outcome always predictable? Transition is:

> a gradual, continuous process of change where the structural character of a society (or a complex sub-system of society) transforms [from one state of equilibrium to another]. Moreover, transitions are not uniform, and the transition process is not pre-determined: there are large differences in the scale of change and the period over which it occurs in any given city. Transitions involve a range of possible development paths whose direction, scale and speed public policy can influence, but never entirely control (Rotmans et al., 2001).

Moreover, while at the level of, say, an entire city, response to new trends, innovations, and developments may appear to be slow and cumbersome, individual entities (for example, entrepreneurs, companies, local governments) may themselves undergo radical change. Certain innovations in technology, behavior, policy and institutions may undergo a breakthrough, and become a dominant design around which learning processes take place, a support or infrastructure base takes shape, and dramatic adoption or "take off" occurs (Rotmans et al., 2001). We shall see examples of this phenomenon later on.

TRADITIONAL WATER MANAGEMENT AND POLICY

Now we are ready to discuss the technological, political, and managerial characteristics of phases of urban water management – and evolution/transitions between them. Each phase is defined by new knowledge and technological innovation, as well as socio-political transitions resulting in policy innovation/reforms. It is also characterized by a unique hydro-social contract displayed in ways that indicate how portions of previous systems remained intact, and influenced current systems.

The Water Supply City/the Hydraulic Era

As we saw in Chapter 2, dependable water supply was required early on to support thriving urban centers and hydraulic engineering of varying levels of sophistication evolved to create water provision systems for these centers' populations. By the mid-nineteenth century, large, centralized, hydraulic engineering systems were introduced to provide secure water supply (Brown et al., 2009). In the UK, Europe, North America, and eventually Australia, disposal of human and animal waste became a high-priority public concern. Concerns with foul, standing water that generated noxious odors, occasional outbreaks of cholera, typhoid, yellow fever and other infectious diseases as waste collected in street gutters, cesspools, and streams (as we discussed) led to public outcry for innovation in developing sanitary sewer systems to shuttle waste (and wastewater) out of cities such as London, Paris, and New York through sanitary and combined sewer systems that discharged waste to local water bodies (Koeppel, 2000).

Major changes in water politics also occurred, including efforts to promote public as well as private investment in the design and development of reticulated sewer systems and improvements to waste transport (Brown et al., 2009). These, in turn, hastened increased public oversight of sewage systems and consolidation of water services, in some cases transforming ownership of supply from private, investor-owned enterprises to public ones. This led to the emergence of a true "hydraulic era" in cities.

Cognitive or attitudinal elements in the hydraulic era included the knowledge to plan, develop, and construct massive water transmission systems. Reflecting back to our discussion in Chapter 2, the growth of cities in antiquity was based largely on an assumption that water supplies were potentially limitless (if externally divertible) and ought to be delivered by the government via large public works projects.

In the case of Australia – which was only colonized by Great Britain in the late eighteenth century (1788), water supply developed in the public psyche as a public right, and among "free" Australians – those released from indentured servitude as prisoners, as well as other early free-born settlers – it was expected that costs should remain low to ensure equitable access by the poor. Moreover, the era of water supply also placed a major emphasis on water rights and the means and legalities by which water is distributed. As a former colony of England, Australia followed the legal system of English Common Law (Common Law). Under Common Law, a system of riparian water rights is dominant; this is not the case in California (as recently as 2013, proposed legislation categorizing water as a human right was defeated in the California legislature, whereas it is a human right in Australia), or elsewhere in the United States. In Australia, newly developed regional governments (later evolving into metropolitan water bonds) levied a centralized tax to fund new infrastructure projects and delivery systems.

California officially adopted English Common Law in 1850, providing an entry point for introduction of the system of riparian rights. However, the gold rush of the 1840s attracted thousands of migrants to California and a system of appropriative water rights developed in a renegade fashion. Under the appropriative system, the first person to take or divert water from a waterway and make use of it, may take as much as needed as long as it is being put to reasonable use. This established the "first in time, first in right" principle. Those who were able to divert water first are considered senior water holders, while those downstream, and consequently at the mercy of the diversions of upstream water rights holders, are labeled junior rights holders. California's geography and patterns of precipitation favored the appropriative system.

Scarcity and the quest for overcoming its constraints was a principal driver of Southern California's – and Los Angeles' preoccupation with water rights. From the inception of Spanish settlement, improvised public works were introduced to harness intermittent stream-flow. Spanish settlers in the southwest, for instance, utilized *zanjas* – literally, hand-tooled ditches – to divert water from streams and rivers to irrigate small farms and orchards, as well as large estates, pueblos, and missions. A longstanding practice in much of Spanish-settled western North America, this improvisation was continued by Anglo settlers who adopted their own variants, appending to these ditches water wheels, mills, and other engineered features (Hundley, 2001). Means to harvest groundwater, arising later, also used improvised advances (Bryner and Purcell, 2003), although connectivity of ground- and surface water resources was not always widely recognized by water users, or by decision-makers.

As we discussed in Chapter 2, Los Angeles not only developed an extensive local system of water supply (*zanja*), but city officials sought ratification of a so-called "Pueblo" water right: an entitlement under traditional Spanish law to lay claim to all needed waters in the vicinity. Later efforts to divert water from the Owens Valley and to develop on a regional basis an emphasis on "appropriative rights" took as their points of departure this tradition, which some have recently characterized as a path-dependent legal process which, while beginning during the era of the state's famed "Gold Rush," continued as the region's economy gradually evolved toward other extractive industries such as agriculture (Kanazawa, 2015). In general, we might say that water law in the region emphasized beneficial off-stream use, not locking up water.[1]

The Sewered City

The sewered city emerged in the mid-nineteenth century – first, in London, England, then throughout Europe, North America, and Australia in the late nineteenth century. At the time, human waste disposal systems included cesspools or outdoor disposal, often in the streets. It is important to bear in mind a critical distinction, however, between sewage collection and sewage treatment. While the latter was widely adopted in cities by the mid-nineteenth century, sewage treatment systems were relatively slow to be adopted. Cities (as far back as Ancient Rome) understood the imperative of moving waste away from the urban populations and, thus, human contact. However, concerns with – and alternative technologies to avoid – dumping wastes into nearby water bodies, sometimes upstream of a city's drinking water intakes, were not properly addressed until the nineteenth century.

The advent of outbreaks of cholera and typhoid across mainland Europe, the U.K., Australia, and the U.S., which sparked public outcry, led to innovation in developing sanitary sewer systems to shuttle waste (and eventually wastewater) out of the city through sanitary and combined sewer systems that discharged waste to local water sources. Cognitive aspects of this transition involve the development of reticulated sewerage systems and engineering knowledge of materials and waste transport.

Early sewerage designs that started in London quickly spread elsewhere. London systems typically combined sewage and stormwater systems, but Australian precipitation/weather patterns are more variable and intense than in the U.K. The cost to build a combined system to handle such intense and large volumes of water was prohibitive, so Australian cities, as well as many U.S. cities, built separate sewerage and stormwater systems. Fresh water also became the norm for transporting human waste through sewered systems. In Australia, the regulatory elements of the sewered city were realized through the development of water boards that levied taxes to support water and sewerage infrastructure. Melbourne's experience is a provocative and telling example of this phase and how it evolved (see: Culture Victoria, 2016).

In the 1880s, the city was nicknamed "Marvelous Smellbourne" because despite the construction of the Yan Yean Reservoir in the 1850s to ensure adequate fresh water, the city (now nearly fifty years old and in the throes of the late Gold Rush) still had no sewerage system (see: Barker et al., 2011). An appalling stench wafted from the many cesspits and open drains. "Nightsoil" (that is, human waste) polluted streets and ran uninterrupted into the Yarra, while "nightsoil" collectors frequently dumped their loads on public roads. Hundreds died in an outbreak of typhoid in the 1870s – a mortality rate 300–400 percent higher than those experienced in comparable U.K. cities, including Glasgow (Barker et al., 2011).

In the 1890s, change occurred, first through creation of a Metropolitan Board of Works (MMBW) (1891) that was charged with building an underground drainage system linked to a pumping station at Spotswood, located on the western banks of the mouth of the Yarra River. Sewage flowed by gravity to Spotswood, and was then pumped to the Werribee Treatment Farm; the Spotswood Pumping Station went online in 1897. At the pumping station, steam engines (later electrical) pumped sewage up a rising main to join the major sewer outfall at the head of the pumping mains where they were then carried to the Werribee Treatment Farm (completed and dedicated in 1898) where it was purified and discharged into the sea.[2] Following the first connection of homes to Melbourne's sewerage system in 1897, there was a steady decline in typhoid mortality. By 1915, 138 108 households were connected and the cases of and deaths from typhoid fever had declined by more than 90 percent (Barker et al., 2011).

Los Angeles operates and maintains the largest wastewater collection system in the United States. This is surprising in that Los Angeles is the nation's second most populous city; however, it is spatially one of the largest, at some 600 square miles in service area. The region served includes Los Angeles and 29 contracting cities and agencies. The City's more than 6700 miles of public sewers convey about 400 million gallons per day of flow from residences and businesses to the city's four wastewater treatment and water reclamation plants. Information on how the city manages its vast sewer system may be obtained in the Sewer System Management Plan (SSMP).

While most sewers are in good working order, some are being repaired or replaced, as part of the 10-year Los Angeles Sewers Program and the Collection System Settlement Agreement (CSSA) legal agreement that defines the maintenance and construction projects and schedules. These include sewers that are very old, seriously deteriorating, or too small for the area they serve. The city has developed rigorous maintenance and construction programs to keep its sewers in reliable condition (Los Angeles Department of Public Works, 2012). The recent drought has also placed additional stress on the system. Household water use reductions have, ironically, led to less wastewater flowing through sewer pipes, allowing solid wastes to collect and eventually to damage pipes (*LA Times*, 2015).

The system's history is dramatic. In the late 1800s, wastewater from Pueblo de Nuestra Señora la Reina de Los Angeles was conveyed through natural waterways to the ocean. In 1892, the city purchased 200 acres of oceanfront property and from 1894 until 1925 raw sewage was discharged into near-shore ocean waters at Hyperion's future location. Visitors to local beaches objected to raw sewage in their recreational waters and in response, the City of Los Angeles built and started operating the first treatment facility at the Hyperion site in 1925: a simple screening plant with an ocean outfall. This plant remained in operation until 1950.

Just after the end of World War II, the city began to develop plans for a full secondary treatment plant at the Hyperion site. When the new Hyperion Treatment Plant opened in 1950, it included a full secondary treatment system and biosolids processing to produce a heat-dried fertilizer. It was among the first facilities in the world to capture energy from biogas by operating anaerobic digesters, which have yielded a fuel gas by-product for over 50 years. At the time, Hyperion was the first large secondary treatment plant on the West Coast, and one of the most modern facilities in the world.

In the 1950s, when the city's population grew dramatically, Hyperion's treatment levels were cut back. By 1957, the new plant was discharging a blend of secondary and primary effluent through a five mile ocean outfall. Hyperion also stopped its biosolids-to-fertilizer program and began discharging digested sludge into Santa Monica Bay through a separate, seven-mile ocean outfall. In recent years, the city has sought to undertake further improvements at Hyperion to keep the plant on the leading edge of environmental protection. Air emission controls continue to represent the leading edge of technology. Odor management facilities are integrated in all improvements. Resource recovery programs capitalize upon every possible opportunity to recycle renewable resources of wastewater and sludge treatment by-products.

The Drained City

The Drained City is described by Brown et al. (2008) as having emerged in Australia in the early twentieth century, reaching its peak in the post-World War II era. As consumer and industrial-related materialism emerged as a global cultural phenomenon, this translated into the twin American and Australian dreams of a single family home with a big yard and a car. Cities developed extensive suburbs: Melbourne capped its growth with imposition of an urban growth boundary (whereas Southern California developed into its famous "sprawl" pattern of low density housing – all of which needed extensive infrastructure to service water and sewage). Furthermore, housing developments were built on and around flood plains, meaning that flood control technology and infrastructure was necessary in residential neighborhoods. Interior waterways (for example, rivers and creeks) were channelized and impervious surfaces characteristic of urban centers crept into suburbs.

Suburbanization and the proliferation of non-porous pavements substantially affected hydrology and impacted local waterways by increasing discharges and the intensity of stormwater from rain events, altering the water quality of surface runoff. Stormwater was viewed as a nuisance. Methods of flood control adopted in this phase were characterized by an emphasis on cost-effective infrastructure, and they were implemented with little appreciation for receiving water's ability to absorb contaminants. Governments emerging from the Great Depression began to spend more money on infrastructure and from the Australian perspective this led to enhanced social welfare. Politically, expansion of suburbs further contributed to a fragmented water governance system in both Australia and the U.S. In California especially, a system of water rights, which had first taken shape during the Gold Rush era, allowed California water governance to evolve more organically and in a decentralized and un-orchestrated manner.

In Victoria, a 1989 Water Act established ten catchment management authorities, including Melbourne Water, which acts as the catchment/ waterway management authority for the Port Phillip & Westernport management area, in addition to being the metropolitan region's bulk water supplier and sewage collection and treatment provider. Each catchment authority is authorized to coordinate and integrate plans for water, land, and biodiversity (VWMS, 1994). Further legislative action (the 2002 Smart Water Fund, established by the Victorian Water Industry as a key part of the industry's response to the ongoing challenges posed by climate change and water scarcity) led to further policy change.

Over the last decade the Fund has invested $50M in over 230 innovative projects across eight research themes. In recent months, the Smart Water Fund Board has undertaken a strategic review of the organization to identify the next phase of innovation and knowledge sharing for Victoria's water industry. The outcome of this review has "recognized the need to imbed the pursuit of innovation within the water businesses and ensure the continual sharing of knowledge" (Smart Water Fund, 2002).

The Waterways City

The Waterways City is representative of the new environmental consciousness that emerged during the international environmental movement in the 1960s. Society's normative values regarding the environment fundamentally changed during this period to incorporate new understandings of limits to growth and the capacity of complex environmental systems to absorb pollution and the rapid volume of water escaping hardscaped cities (Gleick, 2000). Society developed concerns over the impoundment, extraction and ecological health of waterways. According to Brown et al. (2008), this conception of an urban water system is far from mainstreamed. Planning and managing urban water systems that take the value of environmental services into account is becoming more common, but is not standardized in practice.

The first major regulatory measures reflecting this environmental consciousness emerged from Australia and the U.S. with the passage of, respectively, a series of Environmental Protection Acts and the Clean Water Act in 1972. In general, both Australia and the U.S. have taken extensive measures to mitigate the pollutant loads on waterways and to control water quality by regulating wastewater treatment, point and non-point sources of pollution. In this phase, water management became a more expanded area of activity by transcending the realms of water supply planning and provision and becoming an integral element of broader community planning (for example planning for park space and social amenities). Changes to the cognitive elements of water management also occurred during this phase to reflect the scarcity of water resources: more effective, integrative management forms in the form of integrated water resource management in California and integrated and total water cycle management in Victoria, Australia.

In the United States, the Clean Water Act (U.S. EPA, 1972) establishes the water quality regulatory framework that controls pollutant discharges to navigable waters and regulates quality standards for surface waters. The Act sets national wastewater treatment standards, provides grants through the Clean Water State Revolving Fund, and sets standards for contaminant levels in surface waters. The Clean Water Act specifically prohibits point-source discharge of polluted effluent to navigable waters, unless the discharger obtains a permit to pollute. The National Pollutant Discharge Elimination System (NPDES) was established by the Clean Water Act as a vehicle for issuing discharge permits directly, by the Environmental Protection Agency (EPA), or by states having EPA-approved programs.

The overriding idea is to apply technology to ensure that point-source dischargers decrease water pollution emissions. Not only must permits incorporate applicable effluent limitations, but the CWA stipulates that monitoring programs must be implemented by dischargers to ensure compliance. NPDES applies to multiple sectors, including industrial, commercial, and construction sites, and to municipal separate storm sewer systems, so-called "MS4s" (U.S. EPA, 2012; 2016). In the U.S., most states are authorized to allocate discharge permits, provided these permits fully comply with EPA standards for enforcement and monitoring (U.S. EPA, 2012; 2016). In 1987, Congress incorporated Section 319 into the Clean Water Act, which among other things, provides grant funding for non-point-source pollution clean-up programs.

The introduction of the NPDES permit program in 1972, along with various updates to the existing policy, is responsible for many of the significant improvements in water quality in the U.S. (U.S. EPA, 2012; 2016). The Clean Water Act is the amended version of the existing Federal Water Pollution Control Act (1948), which lacked a permit and reinforcement function.

The possibility that urban waterways can be something other than navigation channels or floodways, and that – restored to something resembling their natural condition – they can serve as both recreational amenities and focal points of community engagement has seized the imagination of environmental equity advocates in many cities, including Los Angeles and Melbourne. Moreover, local officials are embracing the view that restored rivers can be a source for urban health. Rivers are increasingly being viewed as focal points for defining a sense of community, and as reminders of the "primal force of clean, flowing water" (Gumprecht, 2005).

The "waterways" movement in Los Angeles and Melbourne is worth comparing, as it may yield lessons for other cities. In both cities, restoration of the Yarra and Los Angeles rivers – the major streams traversing these urban centers – has marked and underscored a debate between using restoration as a means of providing open space and public parkland on the one hand, versus reinvigorating local economies – and fostering gentrification of neighborhoods – on the other. In Los Angeles, the question among restoration groups such as the *Friends of the Los Angeles River* and their allies is: on whose behalf should restoration occur? Will the river become a civic amenity available to all residents, or a gentrified landscape accessible to only a few (Kibel, 2007: 4–5)?

The 51-mile long Los Angeles River was once a rustic, meandering stream serving as a habitat for cottonwood trees and cougars. From 1781, when Los Angeles was founded, until 1913, the river also served as an urban oasis: a picturesque source of the city's water until completion of the Owens Valley aqueduct. Sudden downpours often caused the river to wash over its banks, ravaging homes, vehicles, bridges, and roads, and causing terrible loss of life. Ironically, such seemingly erratic behavior was natural for this modest, meandering stream. It was human intrusion – in the form of dense urban development along its banks – which transformed overtopping into a serious issue.

In the 1930s, as the city became a major metropolis, powerful civic groups sought to promote real estate development along its banks. Through federal appropriations, the Army Corps of Engineers confined parts of the river to a concrete-lined channel that conveyed water to the ocean so efficiently that it remained dry most of the year. While an efficient water carrier, the Los Angeles River became a local eyesore and a symbol of declining civic pride in the city's colorful past (Gumprecht, 2005; 2001).

Restoration efforts are now focusing on removal, ultimately, of some 32 miles of concrete banks, the replanting of native vegetation, and development of greenways and riverfront parks complete with historic markers. Powerful community development advocates favor gentrifying stretches of the riverfront, encouraging high-end, expensive condos to be built at the exclusion of more affordable housing and the working class families that depend on it.

The Yarra, by contrast, still provides some 70 percent of Melbourne's water supply – though as a river, it too underwent massive transformation over time. Rising east of Melbourne near Mt Baw Baw, and flowing some 150 miles to Port Phillip Bay, the stream is home to hundreds of different plants and animals, including platypus, koalas, lyrebirds and native fish. It also remains the subject of massive pollutant inflows, while its wetlands, floodplains and banks are damaged by weeds, and continue to shrink from the pressure of urban growth. Many wildlife species are now endangered. State Government research shows that only 36 percent of the Yarra and tributaries are in good ecological condition (Yarra Riverkeepers, 2015a). Beginning in September 2011, a program to enhance environmental flows on the Yarra was initiated to help restore some of the natural flow patterns. However, severe gaps in historical flow patterns persist. The natural quantity, levels, and timing of flows have not been altered (Yarra Riverkeepers, 2015b). A landmark study (Walsh, 2007) identified three barriers to ecological restoration: (1) a complex problem of ownership along the Yarra inhibits introduction of water-sensitive urban design features effectively (that is, the presence of strong economic disincentives that discourage Melbourne Water from using their stream management budget in the basin); (2) the lack of expenditure on biofiltration systems on publicly controlled (that is, council) lands; (3) a short-sighted focus on equity that seeks to spend monies across the city – rather than to prioritize spending for stormwater management in critical stretches of the Yarra basin.

The Water Cycle City: An Era of Reconciliation

Integrated water resource management (IWRM) and total water cycle management (TWCM) are two water management approaches that have emerged in the United States and Australia, respectively, to address the issues mentioned above. The appearance of each respective water management approach has been accompanied by a set of water policies/legislative changes, social learning, and other supporting institutions. The transition to these management systems has also spurred innovations in technology and policy.

The Australian conception of TWCM is encompassed, for the most part, in an aggressive program of water reuse. According to Brown et al. (2008), limited availability of water has prompted reuse of different qualities of water resources. Conceptually, wastewater would become a component of a closed loop management system having as its goal maximum water reuse in urban centers. This approach would lessen reliance on freshwater taken from streams, creeks and rivers, which serve as environmental flow and support ecological health and diversity.

Brown and Keith (2008) also discuss a new paradigm in water management supportive of more efficient water use behavior in Australia, mindful of energy consumption and nutrient cycles, and the use of a variety of water supplies that are fit for purpose (for example, stormwater, rainwater, recycled water). Brown et al. (2008) describe that the hydro-social contracts established over previous urban water management approaches, where the government would deliver relatively risk-free water, will be challenged now that water managers are asking citizens to accept water of different qualities into their homes and daily routines. Advances and innovations in this area have been supported by a number of water policy reforms over the past two decades and the creation of new governance structures (Office of Living Victoria [OLV])[3] to manage a modern, integrated water system. This era is characterized by an intense interest and focus on innovation and adaptation. This claim has also been amplified by recent research (Parolari et al., 2015) which suggests that, while water-use efficiency in cities has increased over time, the population is growing faster than water supply, necessitating a shift toward wastewater recycling, stormwater capture, and – most importantly – a need for sustainable levels of freshwater use.

The Australian Environment Protection Act (1970) – passed one year after the U.S. National Environmental Policy Act – requires community involvement in environmental decision-making, including local water policy. The Council of Australian Governments Water Reform Framework (1994) enacted water reforms at the national level that were subsequently adopted by states. Reforms seek to promote efficiency and transparency in pricing, trading, water quality programming, institutions, sustainability, and community consultation (Environment Australia, 1994). The water reform framework embraced reduction or elimination of cross-subsidies and making subsidies transparent, clarifying property rights, water allocation to the environment, adopting of trading arrangements in water, and various forms of public consultation (Planning Institute Australia, 2014).

IWRM is used in various places in the U.S. In California, IWRM has taken on a texture and flavor unique to the range of water supply and quality problems characteristic of the nation's most populous state and world's eighth largest economy. Recognizing that the state has one of the world's most heavily managed, coordinated, and intensively developed water management systems to be found anywhere, as well as herculean challenges in supporting agricultural production and burgeoning cities simultaneously, recent IWRM work has fostered a definition more befitting this unique environment. Essentially, IWRM is conceived as the identification of cost-effective and robust options that can best be managed and – if possible – financed locally (which also implies, conjointly, across local jurisdictions), and which combine, in efficient and cost-effective ways, water markets and exchanges, conjunctive (that is, ground- and surface water supply reliance), conservation, and wastewater reuse (Lund et al., 2009).

While IWRM is a farsighted approach to long-term water supply planning, there are a number of practical hurdles that must be surmounted to permit effective adaptation of its principles in an urban context – all of which apply to the theory of transitions discussed in this chapter. First, in general, water supply utilities must adopt new roles and responsibilities, including integration of environmental engineering, public health, financial, ratemaking, and social and economic considerations into the planning process. And, second, the ability to plan effectively is constrained – to some degree – by water utility size; while large investor-owned utilities may well afford to conduct sophisticated needs analyses entailing the screening of feasible alternatives, development of scenarios for demand, and strategy evaluation might not be as feasible for smaller utility districts.

TRANSITION PERIOD AND REFORM IN AUSTRALIA: 1990s AND BEYOND

During the Millennium Drought, Australia adopted a National Water Initiative (NWI) in 2004. The initiative built upon an earlier (1994) national Water Reform Framework. The initiative was directed by a special National Water Commission (NWC 2004–14) whose purpose was to advise the national government on progress in alleviating problems generated by the drought. The overall goal of this initiative was to "drive national water reform" by mandating that states submit water allocation plans to the NWC.

These plans served as the basis for instituting several policies. Most notable of these were: accounting for appropriate water provision for environmental services, managing over-allocated or stressed waterways, introducing registers of water rights and standards for water accounting among users, expanding water rights trading to move water from low- to higher valued uses, improving pricing for water storage and delivery, and meeting and managing urban demand. For its part, the NWI also included commitments to specific actions across several interrelated elements of water policy, ranging from water access entitlements and planning to integrated management of water for environmental and other public benefit outcomes, water resource accounting, urban water reform, capacity building, and community partnerships (Planning Institute Australia, 2014).

While many of the initiatives and methods for their achievement remain in a period of considerable flux, independent observers consider these measures to have:

> (u)shered in a change in agency management: The urban water industry is no longer a series of monopolistic public utilities concerned only with delivery of a product. Competition has been introduced, and policy and regulation have been separated from operations. Regulatory bodies now protect customer interests, oversee the performance of providers, and require providers to show the cost effectiveness of their decisions as well as the effects of these decisions on water prices (Planning Institute Australia, 2014).

CONTEMPORARY VERSUS EMERGING WATER MANAGEMENT AND POLICY

According to Brown et al. (2008), societies may find themselves at many points along the transitional framework, but for the most part, modern, Western cities are still operating from the institutional perspective of the drained and/or waterway city. We face a social, economic, and environmental imperative to transition society toward the water cycle city and ultimately the water-sensitive city. There are some major hurdles. Several academics/authors identify a deficit of policy reform and innovation in the area of urban water (Kiparsky et al., 2013).

All of the conceptual city phases are in existence today and much of the legacy infrastructure and management ideology acts as a barrier to advancing water governance reform and implementation of innovative technologies. Heading into the twenty-first century, several authors have offered their expert opinion as to how modern urban water managers ought to proceed. Gleick (2000) argues that twenty-first century water managers should rethink the previous (pre-water cycle) management paradigm with its focus on large infrastructure projects (for example dams, reservoirs, and aqueducts, so-called "hard path" solutions) and instead focus on solutions that act to simultaneously satisfy/meet social, economic, and environmental needs ("soft-path" solutions).

A Melbournian Future

Melbourne's future challenges loom large over all of this. Between 2010 and 2056, greater Melbourne's population is projected to increase from 4.1 to 6.4 million, with 39 percent of this growth (930 000) occurring by 2026. Of that 930 000, approximately 60 percent of this growth is expected to occur within a 5 km radius of the Central Business District (CBD).

At the same time, our population will age, significantly changing the structure and patterns of demand for housing and services. With existing patterns of water use and supply, demand for potable water in Melbourne could increase from 356 GL to in excess of 534 GL per calendar year, requiring major investment in new supply as early as 2024 (Office of Living Victoria, 2014). Climate variability and natural environmental challenges may deliver extreme shocks (for example, a repeat of the record low inflows to Melbourne's reservoirs in 2006) (Office of Living Victoria, 2014; 2013).

Moreover, urban development will increase stormwater volumes and increase the concentration of pollutants in runoff effluent, presenting major challenges for capturing and managing the quality of this resource, with the dual goal of improving the city's use of "fit-for-purpose" water (for example, for toilet flushing) and improving the health and function of local streams (Askarizadeh et al., 2015).

Community concern over the economic and energy costs associated with implementing large-scale, centralized infrastructure (for example desalination plants) has been a part of the policy landscape during the Millennium Drought and is likely to remain part of the conversation for some time to come. In 2007 Victoria formally committed to building the Wonthaggi Desalinisation Plant, with a maximum capacity of 150 GL/annum, as well as the Sugarloaf Pipeline Project to transport an additional 75GL/annum from the associated Northern Victoria Irrigation Renewal Project to Melbourne (Office of the Premier, 2007). Following widespread public criticism of the desalination project's cost, probable energy consumption, CO_2 generation, and the fact that considerable water savings were generated by conservation, reuse, and other measures, the plant – though built – never went into operation (Mitchell et al., 2008a).

And . . . a Future for Los Angeles

Though it is accepted widely that stormwater is a wasted resource, and various agencies in Southern California are pursuing multi-pronged initiatives that include stormwater capture, water conservation, recycled water, and groundwater remediation, only a portion of stormwater is harvested. The Los Angeles Department of Water and Power (LADWP) currently only captures 27 000 acre-feet per year in centralized spreading grounds for recharging the San Fernando Groundwater Basin. Excess stormwater that cannot be contained in these facilities is discharged to the Pacific Ocean. A paradigm shift is beginning to take place on two levels. First, LADWP began its initial research into a Stormwater Capture Master Plan in the fall of 2013, and produced a plan in 2015 that focused not only on large-scale spreading grounds enhancements but smaller scale green streets, rain

gardens, rain barrels and other "distributed programmatic approaches."
A clearly articulated objective in this plan was to recognize that existing
groundwater pollution or the presence of shallow groundwater could
serve as additional obstacles to using practices that increase groundwater
recharge, as increased infiltration could in some circumstances result in
flooding or mobilization of groundwater pollutant plumes. For example,
portions of the San Fernando and Main San Gabriel groundwater basins
in Los Angeles County are contaminated by such pollutants as the vola-
tile organic compounds trichloroethylene (TCE) and perchloroethylene
(PCE); this complicates efforts both to make use of the basins' resources
and to recharge groundwater supplies.

It should be noted that, by contrast, many of the stormwater capture
projects in Melbourne do not infiltrate the water, but rather "harvest" the
water for above-ground non-potable fit-for-purpose activities, as noted
earlier (Askarizadeh et al., 2015). Interestingly, the use of stormwater for
fit-for-purpose activities (such as toilet flushing) does not appear to be
part of the future visioning for Los Angeles, perhaps due to regulatory and
health concerns associated with the risk of contamination of the drinking
water supply by cross-connections. While such concerns are real, relatively
simple solutions (such as vigorous public education campaigns) have
allowed Victoria to pursue aggressive rain barrel programs, with the result
that it is now routine in Melbourne proper to plumb the house rain barrel
to indoor toilets for flushing.

The second paradigm shift is local government driven. In 2011,
Los Angeles City Council passed a Low Impact Development Ordinance
that is implemented through the city's Department of Public Works.
The law was designed in collaboration with local community groups,
non-governmental organizations, and the local building-trades industry. Moreover, it seeks to encourage redevelopment projects that mitigate runoff pollution and stormwater-induced flood flows by capturing
rainwater at its source via rain barrels, permeable pavement, infiltration
swales, and curb bump-outs to contain water. The immediate benefits of
the program revolve around local water conservation and – in some areas
of the San Fernando Valley – groundwater recharge. Perhaps most durably,
however, the plan seeks to encourage the planting of drought-tolerant
landscaping in residential communities and to provide the efficient means
of providing water to supply this landscaping.

While the city has actually had in place various stormwater regulations
since 1996, the 14 November 2011 Stormwater LID Ordinance requires
stormwater mitigation for a much larger number of development and
redevelopment projects than was previously required. This mandates,
among other things, the inclusion of low-impact development strategies

that incorporate stormwater BMPs in any development or redevelopment project that creates, adds, or replaces 500 square feet or more of impervious area, and it applies to all development and redevelopment in the City of Los Angeles that requires building permits – with very few exceptions (City of Los Angeles, 2011). A final note is in order here: in no case do any of these LID elements allow for indoor plumbing or indoor use of the captured stormwater. This is in stark contrast to Melbourne and, as a design omission, may limit the widespread use of this approach in Los Angeles. We explore these issues in greater detail in subsequent chapters.

NOTES

1. An 1853 California Supreme Court decision (*Irwin* v. *Philips*) ruled that senior appropriators could divert water to the detriment of downstream riparians, based on the rationale that prior appropriation (already in operation) led to installation of "costly artificial works" to sustain appropriators' rights. An 1882 Colorado Supreme Court case (*Coffin* v. *Left Hand Ditch Company*) also upheld the prior appropriation's seniority system (Clayton, 2009: 21).
2. Werribee was not merely Victoria's first "sewage farm." Local annals describe it as a source of civic pride and as the Board of Public Works' most important project "and one of the largest public works undertaken in Australia in the nineteenth century." Simultaneous implementation of collection and treatment by land filtration appears to have been unique in the history of modern cities. Land treatment at Werribee farm consisted of flooding the sewage over open paddocks to a depth of 10 cm, and then allowing the sewage to filter through the soil and ultimately to an earthen drain at the end of the paddock (Barker et al., 2011).
3. The Office of Living Victoria was disbanded after a change in government in 2014, ostensibly due to mismanagement of taxpayer dollars (*The Age*, 2014).

PART II

Traditional and Contemporary Approaches to Water Management and Policy Innovation

Societies across the globe face similar challenges in managing urban water systems defined by social, political, and ecological complexity, risk and uncertainty. However, they also have dissimilar and often unique vulnerability "fingerprints" due to varying water endowment, demands, infrastructure, and institutions (Padowski et al., 2015). While climate change threatens the supply and consequently the availability of water resources by increasing the variability of precipitation and intensity of weather events, and by accelerating the process of aridification in certain regions already encountering water stress, the impacts on countries – and especially their cities – will not be equally felt.

The global phenomenon of urbanization and its associated challenges of population increases, demographic shifts, and changes in water demand and popular expectations further complicate urban water management and water policymaking. In addition to the challenges presented by these global phenomena, policymakers and water managers seeking to adapt and innovate must also contend with a legacy of environmental policies, water rights laws, urban water management approaches, and the resulting institutional arrangements and infrastructure (for example, physical water supply, sewer, stormwater, and flood control systems) resulting from previous eras.

The next two chapters consider two interrelated issues: (1) the relationship between water and energy in large cities (what is sometimes referred to as the water–energy nexus: Chapter 5); and (2) how cities value water – with an emphasis on the economic valuation of water use: Chapter 6. As we have seen, transitioning to more sustainable, safe, and resilient urban water systems will require current and future generations of policymakers, water managers, and society, in general, to confront and renegotiate the implicit and explicit hydro-social contracts and relationships established with water systems from previous eras.

While much can be gained from exploring the nature, mechanism, and contextual factors of these major transitions in water policy and management (in particular, the use of analytical tools derived from social science and policy analysis provide an organized framework for investigating structural changes and transitional pathways from one management era to the next; Brown et al., 2008), fundamental questions of water and its connections to energy, health, and the environment remain to be understood.

Findings from such an analysis may provide promising insights into ways of developing and implementing the innovative and effective water legislation and management regimes needed to combat existing challenges of urban water systems (for example, myopic and fragmented management, policy inaction, and aging infrastructure). Additionally, an analysis of traditional and contemporary urban water management systems may provide tools for handling the increasing complexity and competition between social, economic, and environmental needs, as well as the uncertainty of socio-technical-ecological systems.

5. The water–energy footprint of large cities: productivity and transitional development[*]

Addressing threats to human water security will require getting the most out of locally available water resources. But what does this mean in practice? One way to evaluate water use is to consider its *productivity*, defined as the value of goods and services produced per unit of water used. By improving productivity, communities enjoy the same goods and services, generate less wastewater, and leave more freshwater in streams, rivers, lakes, and coastal estuaries to support biodiversity. Because less water is harvested, treated, and transported, fossil fuel consumption and greenhouse gas emissions are reduced and the water–energy footprint of cities is made smaller.

Although water productivity has steadily improved in the U.S. since the mid-1970s, additional gains are possible here and elsewhere (Gleick, 2003). In this chapter we focus on three general strategies for improving water productivity (see Figure 5.1): substituting higher-quality water with lower-quality water where appropriate, regenerating higher-quality water from lower-quality water by treatment, and reducing the volume of higher-quality water used to generate goods and services.

WHAT ARE THE OPPORTUNITIES FOR SUBSTITUTING?

Many municipal, industrial, and agricultural demands can be satisfied using lower-quality water. For example, treated domestic wastewater that would not be suitable for municipal water supplies may be perfectly suitable for crop irrigation, industrial cooling, and landscape irrigation, to name a few (U.S. EPA, 2004). Although use of treated wastewater in the U.S. is currently limited (< 5 percent of municipal supply), it could be expanded to 17 TL per year (or ~ 27 percent of municipal supply), providing a new drought-resistant source of water in coastal areas where treated wastewater is currently discharged to the sea (National Research

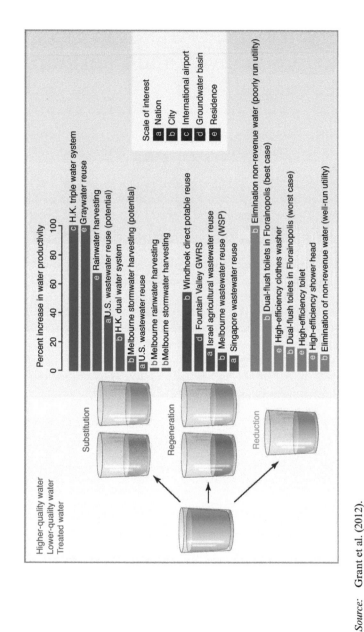

Source: Grant et al. (2012).

Figure 5.1 Three approaches for improving the productivity of higher-quality water

Council, 2012). Large-scale (centralized) wastewater treatment and potable substitution schemes can reduce overall energy consumption and reduce greenhouse gas emissions. In Southern California, substituting potable water with treated wastewater consumes less energy and generates fewer greenhouse gases, compared to inter-basin transfers of water or desalination of seawater or brackish groundwater (Stokes and Horvath, 2009).

Treated domestic wastewater is not the only lower-quality water that can be exploited in potable substitution schemes. Hong Kong's dual water system, which has been in operation for over 50 years, supplies seawater for toilet flushing to 80 percent of its 7 million residents, cutting municipal water use in the city by 20 percent. For example, a triple-water distribution system at Hong Kong's International Airport – consisting of freshwater, seawater and treated graywater from sinks and aircraft wash-down – has helped reduce municipal water use by over 50 percent (Leung et al., 2012).

Potable substitution can also be implemented at neighborhood and single-house scales. Rainwater (from roofs) and graywater (from laundry, dishwashing and bathing) can be used in place of drinking water for a variety of activities. The reuse of graywater for toilet flushing and yard irrigation can cut household municipal water use by 50 percent or more. The energy cost, water savings, and reliability associated with rainwater harvesting depend on engineering considerations (for example, contributing roof area, storage tank volume), local climate, connected end-uses (for example, toilet, laundry, hot water), and temporal patterns (Imteaz et al., 2011). In a case study of a model home in Melbourne (Australia), the use of rainwater tanks to supply water for clothes washing, dishwashing, toilets, and an outside garden reduced household municipal water use by 40 percent (Muthukumaran et al., 2011). However, even in Melbourne, where rainwater harvesting schemes are commonplace, they contribute a modest 5 GL year to the city's overall water budget, which represents 1.2 percent of the city's total water use and 1.4 percent of its municipal supply (Ministerial Advisory Council, 2011).

Storm water runoff from roads and other impermeable surfaces is another locally available source of water, but here the challenge is harvesting and storing the runoff (which can be generated over very short periods of time) and adequately removing contaminants (pathogens, metals, and organic pollutants). These challenges can be overcome through the integration of "natural treatment systems" into the urban landscape, including green roofs, rain gardens, biofilters, and constructed wetlands (Wong et al., 2011) (Figures 5.2 and 5.3). Processes responsible for pollutant removal in natural treatment systems include: gravitational sedimentation of large particles, pathogen removal by solar-UV inactivation and predation, filtration of colloidal contaminants, oxidation of labile organics by

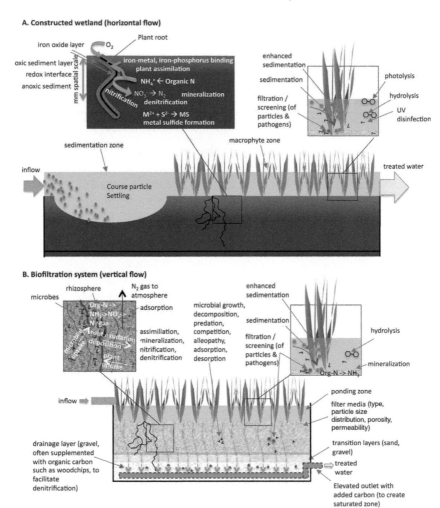

Note: Runoff (e.g., harvested from roofs or roads) enters at the top of the biofilter ("inflow" in the figure) and undergoes treatment by a variety of physico-chemical and biogeochemical processes as it flows downward through sand and gravel layers. The "treated water" accumulates in a drainpipe and from there can be piped into the home for non-potable purposes such as toilet flushing.

Source: From Figure 2 in Grant et al. (2012).

Figure 5.2   Biofilters can be used for treating urban runoff at the house or neighborhood

Note: Shown here are two biofilters designed to treat runoff from a neighborhood (left) and from a suburban street (right). Both are located in Melbourne (Australia).

Source: Photographs by Professor Tim Fletcher (University of Melbourne, Australia) and Dr Brandon Winfrey (UCLA, USA).

Figure 5.3 Biofilters vary in size depending on the volume of runoff they are intended to treat

hydrolysis and sunlight-generated reactive oxygen species, precipitation of metals, and nitrogen removal by bacterially mediated nitrification and denitrification in sediments.

Plants play a key role, taking up excess nutrients and serving both as a source of organic carbon to fuel denitrification, and as a source of oxygen through their root systems to fuel nitrification. As runoff moves through natural treatment systems a portion of the water returns to the atmosphere (evapotranspiration), a portion infiltrates into the subsurface (groundwater recharge), and the rest can be harvested, stored, and ultimately used for non-potable purposes. In Melbourne, stormwater harvesting contributes a relatively minor component (5 GL year^{-1} or 1.4 percent of municipal water use) of the city's water budget (Ministerial Advisory Council, 2011), but including stormwater reuse schemes in new greenfield and brown-field developments until 2050 could result in a seven-fold increase in non-potable water availability for the city (35 GL year^{-1} or 9.8 percent of municipal water use) (Monash University, 2008).

Integrating natural treatment systems into urban landscapes confers many benefits beyond improving human water security. In warmer climates the evapotranspiration of runoff moderates the urban heat island effect (Mitchell et al., 2008b), which can substantially reduce energy requirements for cooling, while infiltration recharges the groundwater and provides environmental water for local wetlands and riparian zones (Wong et al., 2011). The construction of new, or reinvigoration of existing, wetlands creates habitats for resident and migratory species and sustains biodiversity through enhancement of habitat heterogeneity, connectivity and food web support (Murray and Hamilton, 2010). By locally detaining and retaining stormwater throughout the catchment, less runoff enters rivers and streams, pollutant loads are reduced, and flow regimes resemble more closely pre-development conditions (Bunn and Arthington, 2002). As a result, streams are less likely to top their banks and cause flooding and the negative effects of urbanization on stream health and function – collectively known as the "urban stream syndrome" – can be mitigated (Walsh et al., 2005a).

WHAT ARE THE OPPORTUNITIES FOR REGENERATION?

With adequate treatment, higher-quality water can be regenerated from wastewater. Because additional goods and services are produced every time a parcel of water is recycled, regeneration has the potential to significantly increase water productivity. A prime example of regeneration is potable reuse, in which wastewater is treated using conventional and advanced methods, and then added back to the water supply either directly ("direct potable reuse") or indirectly by holding the water for a time in groundwater or surface water reservoirs ("indirect potable reuse") (U.S. EPA, 2004).

Apart from a few small-scale facilities, direct potable reuse is not practiced in the U.S. However, several indirect potable reuse facilities are operational. The world's largest is the Groundwater Replenishment System (GWRS) in Fountain Valley (California), which treats up to 97 GL year^{-1} of domestic wastewater using conventional (primary and secondary sewage treatment) and advanced (microfiltration, reverse osmosis, and UV disinfection) techniques (Leverenz et al., 2011). Water produced by the GWRS provides approximately 20 percent of the water needed to maintain the local groundwater aquifer in Orange County, a primary source of municipal supply for more than two million residents.

Internationally, the longest running example of direct potable reuse is in Windhoek (Namibia), where recycled wastewater (mostly domestic sewage) has been added to the potable water distribution system more or less continuously since the late 1960s with no obvious adverse health effects among the population of several hundred thousand (du Pisani, 2006). The current facility produces enough water (some 7.7 GL year^{-1}) to meet approximately 35 percent of the city's municipal water needs.

Among centralized options for augmenting potable water supplies, potable reuse is preferable to inter-basin water transfers for several reasons (Schroeder et al., 2012): (1) inter-basin water transfers reduce the water available at the source for critical ecosystems and agricultural production; (2) transporting water over long distances can be energy and carbon footprint intensive; and (3) the water transmission systems are vulnerable to disruption by natural and man-made disasters, such as earthquakes and acts of terrorism. All three problems are evident in California, where the southern part of the state has long relied on water imported from sources located hundreds of kilometers to the east and north.

In 2001, an estimated 4 percent of the electric power consumption in California was used for water supply and treatment (largely transportation) for urban and agricultural users; this estimate increases to 7 percent if end uses in agriculture (which are mainly related to pumping) are included (California Energy Commission, 2006). The depletion of source waters in the state has led to habitat deterioration, the decline and extinction of native fish species, the near collapse of the Sacramento–San Joaquin River Delta ecosystem, and the desiccation of Owens Lake, whose dry lake bed is arguably the single largest source of asthma- and cancer-inducing respirable suspended particles in the U.S. (Reheis, 1997). Potable reuse also has advantages relative to the desalination of seawater. By one estimate, potable reuse consumes less than one-half the energy (c. 1000 to 1500 kWh ML^{-1} beyond conventional treatment) required for the desalination of seawater (c. 3400 to 4000 kWh ML^{-1}) (Schroeder et al., 2012).

Relative to the classification scheme presented in Figure 5.1, some non-potable wastewater reuse is best described as regeneration, provided the effluent replaces water of an equal or lower quality, such as river diversions. For example, 73 percent of Israel's municipal sewage is treated and reused for agricultural irrigation, equal to roughly 5 percent of the country's total water use and 13 percent of its municipal supply. In Singapore, 27 GL year^{-1} of highly treated domestic wastewater is used primarily for industrial applications, equal to 5 percent of total water use and 9 percent of municipal supply (Schnoor, 2009).

Relatively low-energy centralized approaches for non-potable wastewater reuse are also available, such as waste stabilization ponds (WSPs) in which sewage is directed through a series of open-air shallow ponds where physical processes (flocculation and gravitational sedimentation), microbial processes (algal growth, aerobic and anaerobic heterotrophic metabolism, nitrification, denitrification) and exposure to sunlight jointly remove pathogens, organic contaminants, and nitrogen (Varon and Mara, 2004). Effluent from WSPs can irrigate crops (Figure 5.2) or recharge groundwater aquifers, and the ponds themselves may provide a much-needed quasi-wetland habitat for waterbird conservation. The world's largest WSP system, the Western Treatment Plant in Melbourne, Australia, produces 40 GL year^{-1} of treated wastewater, equivalent to 11 percent of Melbourne's municipal supply, and uses approximately 500 kWh ML^{-1} less energy compared to conventional wastewater treatment (Saliba and Gan, 2012). Recycled water from the Western Treatment Plant is used for a variety of non-potable applications, including in-plant uses and dual pipe schemes for the irrigation of agricultural crops, gardens, golf courses, and conservation areas.

Primary concerns associated with wastewater reuse include the build-up of contaminants and salts in soils (in the case of wastewater irrigation), and the possibility that incomplete removal of chemical or microbiological hazards during treatment may cause disease in an exposed population. Disease risk can be evaluated on a case-by-case basis using a statistical framework, such as quantitative microbial risk assessment, that predicts a population's disease burden given the types and concentration of pathogens likely present in the water and particular exposure scenarios (World Health Organization, 2006).

WHAT ARE THE OPPORTUNITIES FOR REDUCTION?

Water productivity can also be improved by reducing the volume of water used to produce a fixed value of goods and services. A modeling study of the water supply system in Florianopolis, Brazil concluded that replacing single-flush toilets with dual-flush toilets would reduce municipal water use in the city by 14–28 percent, and reduce energy use at upstream (drinking water) and downstream (wastewater) treatment plants by 4 GWh per year – enough energy to supply 1000 additional households (Proenca et al., 2011). An analysis of 96 owner-occupied single-family homes in California, Washington, and Florida concluded that installation of high efficiency showerheads, toilets, and clothes washers reduced household use of municipal water by 10.9, 13.3, and 14.5 percent, respectively. Because water is not technically required for bathroom waste disposal, installation of composting toilets and waterless urinals can reduce municipal water use even further (Anand and Apul, 2011).

Agriculture accounts for the majority of global freshwater withdrawals (Gleick, 2003; UN Water, 2014), and thus even small improvements in water productivity in this sector can result in substantial water savings. Water savings can be achieved by switching to less water-consuming crops, laser-leveling of fields, reducing non-productive evaporation of water from soil or supply canals, changing irrigation scheduling, and adopting more efficient sprinkler systems, including micro-irrigation techniques (drip irrigation and micro-sprinklers) that precisely deliver water to plant roots. These approaches could help mitigate escalating water demand in the west north-central region of the U.S. associated with energy crops, such as corn, if projected increases in U.S. biofuel production are realized (Elcock, 2009).

Drinking water is often lost after it leaves treatment plants due to physical leaks in urban water distribution systems and poor accounting. Worldwide, the total volume of this "non-revenue water" is estimated to be 49 TL per year (Kingdom et al., 2006). Pipeline losses range from over 50 percent in much of the developing world to less than 10 percent in well-run utilities (Kingdom et al., 2006). The World Bank estimates that if just half of the losses in developing countries were eliminated, $1.6 billion would be saved annually in production and pumping costs, and drinking water could be extended to an additional 90 million people without the need for new treatment facilities (Kingdom et al., 2006).

IMPORTANCE OF WATER QUALITY: A FURTHER PRIORITY

Protection of water quality is also a priority. The Catskill Mountains supply drinking water for New York City, as noted earlier (see Chapter 2). When agriculture and residential developments threatened surface water quality, the City considered building an $8 billion water treatment plant, but instead opted to spend $1 billion buying land and restoring habitat in the water supply catchment. This approach obviated the need for a treatment plant, saved the city billions of dollars in capital and ongoing operation-maintenance costs, and preserved a critical ecosystem (Jackson et al., 2001).

What is the Right Mix of Wastewater Reuse and Water Saving Schemes?

The wastewater reuse and water saving schemes described above are each tailored to a particular scale of implementation (from single homes to entire countries), population density (urban to rural), and level of techno-logical sophistication (from high-tech to low-tech) (Figure 5.4A). Potable substitution schemes utilizing advanced wastewater treatment may be feasible in an urban context, but not so in a rural context. Furthermore, no single scheme simultaneously maximizes wastewater reuse, minimizes wastewater generation, and minimizes stormwater runoff (Figure 5.4B). How does a community identify the right mix of schemes that will opti-mize their water systems?

One study evaluated infrastructure options for a hypothetical residen-tial development in the southeast of England, and concluded that every community has a technological state-of-the-art equilibrium beyond which trade-offs are required. Wastewater reuse and water saving schemes can improve water use, energy use and land use up to the equilibrium point. Beyond the equilibrium point, further reductions in water use require increasing either energy use (if high-tech options are utilized) or land use (if low-tech options are utilized) (Makropoulos and Butler, 2010). Human behavior should also be considered in assessments of optimal water man-agement strategies, as was done in an elegant systems modeling study of water supply options in Chennai, India (Srinivasan et al., 2010).

Because end-user behavior affects all aspects of water and wastewater management, changing attitudes and expectations may be more effective than finding infrastructure solutions to water scarcity. Turf grass consumes upwards of 75 percent of residential drinking water in arid and semi-arid areas of the U.S. (Milesi et al., 2005). If water resources in this region continue to dwindle, reducing the volume of water used for yard irrigation

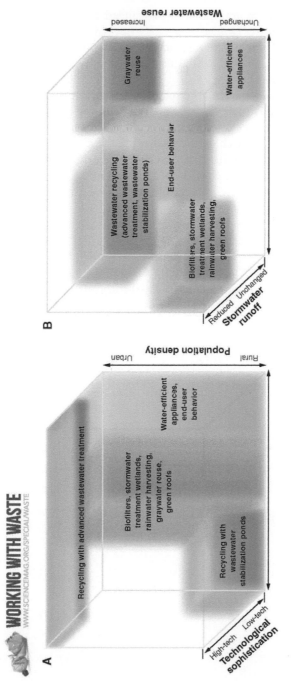

Source: Grant et al. (2012).

Figure 5.4   Wastewater reuse and water saving schemes

– for example, by implementing advanced irrigation technologies – may not be sufficient. Homeowners may have no choice but to replace turf grass with xeric landscaping (Larson et al., 2009). Such wholesale rethinking of our relationship with water is an example of a "soft-path" approach to water management.

As a general principle, the soft-path approach to water management is characterized by (Brooks et al., 2009): (1) viewing water as a service rather than an end in itself; (2) adopting ecological sustainability as a fundamental criterion; (3) matching the quality of water delivered to that needed by the use; (4) planning from the future back to the present; and (5) ensuring community and citizen involvement in water management planning. As we shall see in the next chapter, the efficacy of such an approach partly entails understanding the value of water in cities.

NOTE

* This chapter is largely extracted from Grant et al. (2012).

6.   How cities value water and why it matters: economic and non-economic approaches

Water is valuable. It not only satisfies a basic biological need, but it conjures in all of us an importance, or worth, that goes beyond mere survival, and in many different ways. This is true for all people, whether they live in cities or not. Water is – among other things – a commodity that someone provides to us, at some cost to ourselves and to others. In addition, however, values toward water are shaped by custom, primordial tradition, and even religious faith – all of which infuse water with widely varying symbolic, as well as practical, value (Linton, 2010).

As a result, how we value water – especially in cities – is shaped and determined by what we term *non-market* values. Among the latter, which helps us to understand domestic water use, are such treasured amenities as cleanliness, comfort and convenience, the so-called "three domains of daily life" (Shove, 2003a and 2003b). These non-economic values are important for two reasons, as this chapter will discuss. First, they help explain how urban residents view their connections to water infrastructure, technology and daily practice. In most instances, these connections are invisible to the average person, and thus most of us rarely appreciate how arduous and difficult obtaining and treating freshwater has become in modern cities (Supski and Lindsay, 2013; Giddens, 1984).

Second, these values underscore various patterns of both water and energy consumption, and thus reflect deeper cultural and social habits and attitudes toward the environment. Finally, because these non-economic values reflect deep-seated attitudes, habits and cultural practices, they resist change. As a consequence, when urban water decision-makers, including utilities, seek to promote the use of recycled and/or alternative water supply sources, their actions might be resisted due to a syndrome we may call the "good water, bad water" distinction: the former is water that has been "processed, controlled, commodified," while the latter refers to "untreated metabolized water, to be found in city rivers, lakes, rainwater, sewerage." Moreover, this distinction underlies many cultural misunderstandings and perceptions regarding the actability of these alternative

---

BOX 6.1   SOME NON-ECONOMIC VALUE DOMAINS FOR WATER

- *Comfort* – domestic hot water provision facilitates bathing, washing, showering.
- *Cleanliness* – changes in practice of frequency of washing, bathing, etc. Tied to significant cultural understandings of what it is to be clean.
- *Convenience* – proliferation of technologies, products, arrangements that make everyday life easier (washing machines and showers), but also use resources in ways that are often not sustainable (Shove, 2003b).

Source: Supski and Lindsay (2013).

---

water sources. However, it should also be noted that these culturally formed values (see Box 6.1) can be modified through education, information, and demonstration of the efficacy of alternative approaches (see Kaika, 2005; Head and Muir, 2007; Moy, 2012; Supski and Lindsay, 2013).

There is another element regarding the value of water this chapter will discuss: the role of economic factors in explaining urban water use and behaviors. An important question in this vein is: how much should it *cost* us to obtain suitable amounts of water for various purposes? Another question might be: *how* should the *price* charged for obtaining this water be determined, and by *whom*? Various fees, charges, taxes, and "trading schemes" have long been applied to the management of water supply, as well as to efforts to improve water quality by reducing pollution. Water conservation, particularly in the residential sector in cities, has become an international challenge.

While utilities in traditionally water rich regions have tried to encourage conservation, in part to provide greater flexibility in meeting additional needs, reduce operating costs for water treatment, and lessen burdens on existing infrastructure, in water-supply challenged regions such as the southwest U.S., such efforts have also been driven by chronic and worsening drought and rapid population growth. In both cases, these issues are becoming widely covered in popular media (for example, Doig, 2012), and in the southwestern U.S., this was true even before the advent of our current drought.

Finally, this chapter examines the increasingly contentious issue of privatization in the management of urban water. Private ownership of freshwater provision and treatment is one of the world's most rapidly growing businesses, and especially popular in large cities. Among the most important reasons for its advent is the need for capital to meet growing demands for clean water in developing nations. These demands, and the role of privatization in meeting them, have power, equity, infrastructural, and aspirational implications. The dramatic growth of privatization has prompted questions regarding political accountability, economic fairness, and the willingness of vendors vigilantly to prioritize public health and other community concerns above handsome returns on investment. Critics charge that public participation in decisions regarding pricing, service, and access is sharply limited because policies regarding these issues are established by corporate boards that are advised by business managers – not by politically sensitive elected officials.

The result, especially in developing nations where the capacity to regulate private enterprise is weak, and where global financial entities encourage private investment to reduce public indebtedness, is that public goals regarding water are held in lower regard than profit. In short, decisions over water management are made exclusively, not inclusively. Finally, private providers' operations are less transparent to the general public than is the case with public providers. Once control over local supplies and their provision are turned over to private companies, or even when plans to do so are contemplated, enterprises may shield their operations to protect proprietary interests. When this occurs, local political support for privatization often erodes among public officials. This situation has arisen in recent years in Pune, India; Karachi, Pakistan; Cochabamba, Bolivia; Manila, Philippines, and elsewhere.

HOW DO WE PAY FOR WATER?

Internationally, understanding how people use and value water requires new insights into such factors as cities and urban sub-groups' "cultural histories." These are efforts to account for the role of gender, ethnicity, shared interests and experiences in explaining the complexity of water use (Sofoulis and Williams, 2008; Sofoulis, 2005; 2006; Supski and Lindsay, 2013). This cultural approach acknowledges the importance of the divergence between cultural, social, political, economic and technological frameworks. It also recognizes the overarching role of institutions, including centralized urban utilities that too often promote large-scale engineering projects that are controlled by huge bureaucracies and which wield near-absolute and total responsibility for water supply, quality and disposal.

Among pricing schemes, three approaches predominate among utilities: uniform, decreasing, and increasing bloc rates. While, in general, increasing bloc rates – a method whereby consumers pay more per unit volume of water once their use exceeds a nominal "conservation" threshold – can

Table 6.1   Comparing residential water use and charges in Australia and California

Location	Residential use,** lpcd (gpcd)	Urban use,** lpcd (gpcd)
Portland, OR	219 (58)	390 (103)
Albuquerque, NM	282 (74)	587 (155)
Tucson, AZ	367 (97)	544 (144)
Denver, CO	393 (104)	604 (160)
California	394 (104)	568 (150)
San Francisco	172* (46*)	295* (78*)
Oakland/East Bay	277*–316 (73*–83)	439*–469 (116*–124)
San Diego	277*–350 (73*–92)	490*–524 (129*–138)
San Jose	307–323* (81–85*)	489–519* (129–137*)
Los Angeles	345*–376 (91*–99)	450*–547 (119*–145)
Sacramento	428*–455 (113–120*)	642–667* (170–176*)
Australia	204 (54)	318 (84)
Melbourne	150 (40)	238 (63)
Brisbane	172 (45)	289 (76)
Canberra	191 (50)	288 (76)
Sydney	207 (55)	312 (83)
Perth	284 (75)	399 (108)

Notes:
* From Urban Water Management Plan.
** Does not include distribution system losses.

Source: Cahill and Lund (2011).

induce water savings, considerable variation in this reduction has been documented by resource economists who, over the past 25 years, have undertaken detailed case studies in a variety of community contexts. While there is broad agreement that the overall price of water influences demand, there remains wide debate over the comparative effectiveness of these systems – and what other factors might encourage us to place a high economic value on water (see Table 6.1).

To induce greater residential conservation, water providers have adopted a variety of strategies – alone or in combination – that are intended to drive more efficient water use. These strategies range from incentives to install water-saving landscape to indoor water-appliance standards and special rebates or loan programs to adopt these innovations to various water pricing schemes. In the western U.S., California included, many of these programs were introduced beginning in the 1970s in response to protracted drought. Municipalities adopted metering, incentives for adopting drought-tolerant landscaping, aggressive public outreach programs, and water appliance standards, and also resorted to a number of water-pricing schemes to induce conservation (for example, Hanak et al., 2012b; City of Los Angeles, 2008; Gleick, 2003).

Residential consumption tends to vary as a result of climate, square footage of lawns and general landscape preferences, home size, income, and aggressiveness of public education and outreach (Nieswiadomy and Molina, 1989; Stevens et al., 1992; Nieswiadomy, 1992; Hodgins, 2010). Moreover, rate structures are usually so complex – combining fixed and variable charges in different combinations – that ascertaining exactly how a particular price structure affects consumption is methodologically difficult to determine (Arbues et al., 2003).

In California, for example, a number of economically driven decisions have affected consumption, including rising costs for imported water from the Bay-Delta region in Northern California and the Colorado River, as well as state-mandated water conservation efforts (for example, Senate Bill 7), which mandate reductions in per capita demand by 2020 (Mills et al., 1998; OCWD/OCSD partnership, 2004).[1]

Financial Mechanisms

The price structures used by water utilities can serve the dual objectives of recovering water delivery costs as well as promoting conservation. Metered billing, or usage-based billing, is an alternative to charging customers a flat monthly rate. It leads to a water bill that varies with the volume of water used, sending a price signal to customers. The metered cost of water should reflect the marginal cost of water as a means of reducing demand (AWWA, 2012). Metered billing requires the installation of water meters for customers to measure water use on a per unit basis. Although there are some empirical and theoretical studies that suggest water demand can be price inelastic after a point (Chestnutt and McSpadden, 1991; Dalhuisen et al., 2003; Espey et al., 1997), or that the savings achieved by metered billing can be temporary, metered billing has often been shown to reduce water use and is generally considered a viable mechanism for conservation. Maddaus (2001) reported a 13 percent reduction in demand after Davis, CA changed from a flat monthly fee to a metered billing system, and found in a literature review that metered billing systems generally save 10–30 percent, and sometimes up to 50 percent. In recent years, water utilities have been shifting to Advanced Meter Infrastructure (AMI) or "smart meters" that can record water usage in brief intervals and can communicate data for billing and monitoring. These meters have the potential to detect leaks and provide detailed usage information for furthering conservation and reducing utility costs.

There are different types of metered billing structures. The uniform volumetric rate is a unit rate that remains constant despite the quantity of water used. Although the unit rate remains constant, the total price of water increases with quantity used. A seasonal rate is a unit rate of water that changes depending on the season to reflect the varying cost of water delivery during different times of the year. A tiered or block rate is a unit rate of water that changes as the quantity used increases. An increasing tiered rate, which bills customers at a higher unit price as water use increases, may be utilized to send a strong conservation signal. Tiered rate systems are becoming more common. For example, 27 percent of utilities in California used tiered rates in 1991, while 43 percent used them in 2006 (Black & Veatch, 1995; 2006; Donnelly and Christian-Smith, 2013).

An allocation or budget-based rate is a tiered rate that is based on customers' "allocations" or "budgets," which are calculated using factors such as household size (for indoor use) and landscape area and evapotranspiration (for outdoor use). When a customer's water use exceeds allocation, the unit rate of water is much higher. A benefit of this system is its individualized nature in establishing reasonable need. However, the cost of determining the appropriate allocation for each customer in its district may be expensive for a utility. Upon switching to a new rate plan for promoting conservation, an agency may follow up with programs for assisting customers in reducing water use, such as rebates for water-efficient indoor and outdoor technologies or turf replacement. Residential audit programs may be offered as well.

In the U.S., utility-led efforts to promote reductions in water use generally, and residential uses in particular, have made measureable progress in recent decades. As a region, southern California is often regarded as a model for what conservation programs can achieve through a combination of tiered-pricing, metering, and forward-thinking public outreach. We assess residential water use trends in Orange County, a sub-region of southern California that has witnessed wide adoption of variable rate pricing intended to foster residential conservation.

Despite impressive water use reductions following introduction of these programs, dramatic reductions in use tend to be followed by more modest annual reductions or – in some instances – increases in use. The novelty of tiered-rate programs together with consumer adjustments to the "shock" of their initial introduction generates lower water use. However, once consumers become acclimatized to higher rates and adjusting to using less water, consumption patterns level off. This finding is important for two reasons: (1) further gains in conservation will probably require more aggressive and possibly less popular measures; and (2) these less popular efforts will probably include further increases in the price of water, partly in response to continued infrastructure needs and curtailed imports of water from outside the immediate region. This latter issue is already stirring political debate in California and elsewhere.

THE ECONOMICS OF CONSERVATION

Water conservation, particularly in the residential sector, has become a national challenge. While utilities in traditionally water-rich regions have tried to encourage conservation, in part to provide greater flexibility in meeting additional needs, reduce operating costs for water treatment, and lessen burdens on existing infrastructure, in water-supply challenged regions such as the southwest U.S., such efforts have also been driven by chronic and worsening drought and rapid population growth. In both cases, they are becoming widely covered in popular media (for example, Doig, 2012).

In an effort to induce greater residential conservation, utilities have adopted a variety of strategies – alone or more often in combination – that are intended to drive more efficient water use. These range from incentives to install water-saving landscape to indoor water-appliance standards and special rebates or loan programs to adopt these innovations to various water pricing schemes. In the western U.S., California included, many of these programs were introduced beginning in the 1970s in response to protracted drought. Municipalities adopted metering, incentives for adopting drought-tolerant landscaping, aggressive public outreach programs, and water appliance standards, and also resorted to a number of water-pricing schemes to induce conservation (for example, Hanak et al., 2012b; City of Los Angeles, 2011; Gleick, 2003).

Residential consumption tends to vary as a result of climate, square footage of lawns and general landscape preferences, home size, income, and aggressiveness of public education and outreach (Nieswiadomy and Molina, 1989; Stevens et al., 1992; Nieswiadomy, 1992).

Nationally, efforts to induce conservation in the residential water use sector have resulted in declines of some .44 percent annually (or 4.4 percent per year over the past 10 years) and 13.2 percent since the early 1980s (Hodgins, 2010). In Orange County, California, the focus of this investigation, water use reductions significantly exceed these national trends. In the past decade, for example, declines in total water use have been on the order of 1 percent per year, despite population growth rates also averaging 1 percent per year (Orange County Community Indicators Project, 2011: 66; 68).

In accounting for these trends and their variation, as well as their underlying motivations and prospects for long-term water use reductions, it is clear – in light of resource economist studies – that we must focus on distinctive factors favoring conservation if we are to understand these issues. Nationally, the principal driver of residential water consumption continues to be outdoor water uses, which are affected directly by weather and temperature – wetter soils lead to less use, while hotter temperatures lead to more. Furthermore, outdoor uses tend to increase water use rates by a significant margin in the U.S. (Hodgins, 2010).

Another major policy innovation in Orange County that has been investigated in other studies (Mills et al., 1998). In short, progress made in inducing residential water use reductions in Orange County can be seen as a kind of benchmark against which one can measure the effectiveness of conservation programs driven by extreme drought and changes in policy.

A Case Study: Southern California

Data on major water uses – residential, commercial/industrial, and overall – for the period 2005–10 was collected for 31 water utilities in Orange County, California via a variety of published reports and Internet sites available. In addition, we amassed information on water rate structures and conservation policies adopted by these 31 water districts during the same period. Our overall goal was to track relationships between the various rate structures on the one hand, and increases or declines in water use on the other, as a means of assessing the overall effectiveness of ratepayer-based water conservation programs.

We examined three relationships germane to understanding efforts to reduce urban water use by encouraging residents' water conservation practices. These relationships were as follows. First, we examined the overall impact of tiered rate (or increasing bloc rate) systems on water use for 2010. Second, in order to account for changes in residential use following adoption of tiered rate programs, we examined the relationship between tiered pricing systems in Orange County and residential water use during the period of investigation (2005–10). In these analyses, we assessed water use patterns measured in acre-feet per year, among systems employing some type of tiered rate structure (that is, a structure which charges more for greater volumetric use).

Third, we tried to isolate those single-family residential accounts within the water districts analyzed by focusing on accounts belonging to homes on large tracts in our study region. Because the largest residential water uses are for landscaping and other outdoor usage, we wanted to gauge the degree to which water rate systems affect conservation in these areas. In effect, changes in the actual number of single-family user accounts have a dramatic effect on per capita water use, something we were able to assess by tracking water use relative to declines in single-family accounts.

Findings: Water Use Trends

During the period 2005–10, water use reductions have been significant in Orange County. We identified six major trends, most prominent in a handful of communities, as noted below. First, in many communities, active conservation measures taken by utilities, especially *tiered rate structures*, do account for significant percentage drops in usage. These drops were most significant from 2005–10 in Laguna Beach, Fullerton, and GSWC West Orange (20, 23, and 23 percent respectively). However, such gains have not been seen in most other communities in the county, prompting the question as to what factors may make these cases unique.

We found that three factors actually account for most of these conservation gains: income, population density within a water district, and single-family house tract size. Simply stated, the more modest the income, the higher the population density, and the smaller the tract size of a family's property, the greater the conservation gains. This finding also confirms previous studies that note considerable variation in gallons per capita per day (GCPD) consumption among retail water agencies in Orange County, primarily as the result of outdoor uses associated with larger home lots, micro-climate variation (coastal communities have lower outdoor irrigation needs due to fog and cloud cover), and extent to which new water appliance rebate and other incentive programs have been adopted (Center for Demographic Research, 2009). Three other findings were significant.

First, we found that the overall impact of tiered rate systems on water utility customer consumption, when compared to utilities that did not employ tiered rate structures, is not entirely clear. Somewhat surprisingly, we found that there was no strong correlation between tiered vs. non-tiered rate systems, although there does seem to be a relationship between price and water use in districts where metering is widely employed. The overall difference in water savings between tiered and non-tiered approaches is weak, with a correlation coefficient of .24 for 2010.

Figure 6.1 depicts the relationship between the longevity of tiered rates and residential water usage for 2010, while Figure 6.2 depicts change in water use as a result of longevity of a tiered rate system for the entire period of analysis: 2005–10. While the scatterplots shown in these figures suggest that tiered approaches do lead to residential use reductions, no statistically

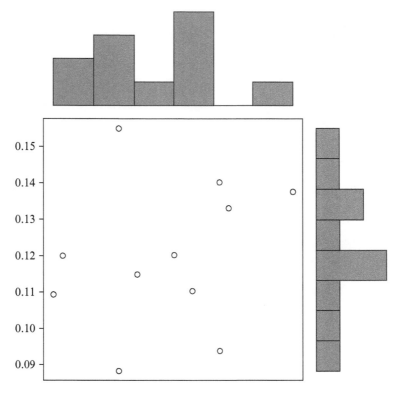

Notes:
P = 0.2581; without outliers, P = 0.6611. Residential change 05–10, tiered vs. non-tiered: P = 0.5857.
Longevity of tiered rates vs. residential usage 2010: correlation = 0.246; determination = 0.061.

Figure 6.1   Longevity of tiered rates vs. residential usage, 2010

significant link between tiered rates and lower usage in the single year 2010, or of a pattern of decreasing usage between 2005 and 2010, were observed. Moreover, overall residential water usage through tiered rates, while declining during the period of study, actually exhibited a diminishing effectiveness over time. In short, the longer tiered rates are employed, the less effective they are in reducing overall residential consumption.

A second important finding is that the longevity of tiered rates does seem to have a fairly strong effect on *single-family residential water use*. This objective is precisely the aim of such programs in the first place. Gauging from our review of urban water management plans for various Orange County communities (Figures 6.3 and 6.4), reductions in single-family

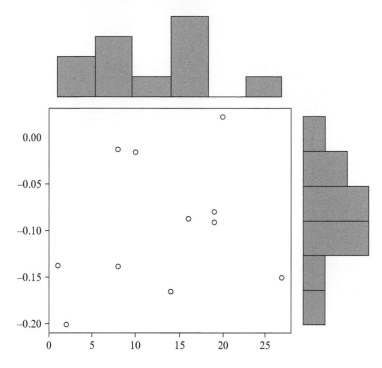

Note: Longevity of tiered rates vs. residential change 2005–10: correlation = 0.238; determination = 0.057.

Figure 6.2 Longevity of tiered rates vs. residential change, 2005–10

home water use are likely due to the fact that tiered rates have their most dramatic effect upon outdoor uses.

And finally, even though residential consumption has declined overall in most communities, there have been marked increases in per capita residential consumption, prompting additional questions regarding what makes these communities decided outliers compared to other water districts in Orange County. These communities and districts include Yorba Linda Water District, whose consumption increased some 13 percent (although single-family homes saw a modest increase in usage), and Santa Ana, where use increased nearly 2 percent. Yorba Linda, a relatively affluent community in the county's northern tier, has experienced an increase in large single-family homes built on fairly large tracts in recent years. By contrast, Santa Ana, one of the county's oldest and most modest-income communities, had among the lowest per capita rate of water consumption in 2005 – with small home lots, lower rates of outdoor use, and thus, little

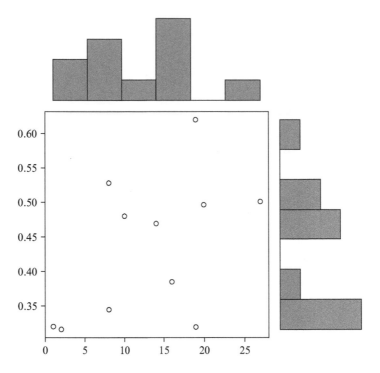

Notes:
SF consumption in AF divided by number of SF accounts (not including Seal Beach and Serrano Water District). SF usage per account 2010 tiered vs. non-tiered: P = 0.3465; without outliers, P = 0.6426.
SF change 05–10, tiered vs. non-tiered: P = 0.3472. Longevity of tiered rates vs. SF usage per account 2010: correlation = 0.506; determination = 0.256.

Figure 6.3   Longevity of tiered rates vs. single-family usage per account, 2010

"cushion" for water use reductions to begin with. Further observations regarding these differences in water use, and their deeper implications, are discussed in the following section.

SUMMARY AND CONCLUSIONS

Five conclusions emerge from this study. First, despite aggressive adoption of tiered rate structures among most water utility districts, we find no statistically significant link between tiered rates and lower usage in 2010, or in explaining decreased usage between 2005 and 2010, if tiered rates are

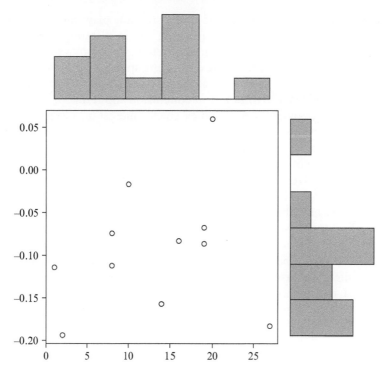

Note: Correlation = 0.187; determination = 0.035.

Figure 6.4   Longevity of tiered rates vs. single-family use changes, 2005–10

taken alone. Second, there is a weak – but observable – inverse relationship between the longevity of tiered rates and 2010 usage and 2005–10 usage drop. In short, greater drops in use are observed for newer tiered programs, but only for a limited period of time. While there appears to be an initial "shock-induced" decline in water use, tiered rates generally decline in overall effectiveness over the long run. As customers become acclimatized to using less water, and paying more for it, the effects of tiered rates upon overall consumption level off. Moreover, this observation is true for both residential and overall per capita consumption.

Third, districts with tiered rates in place, which are designed to discourage consumption, tended to mirror the overall trend of districts throughout the entire county in regards to across-the-board consumption decreases. In short, there were no statistically significant gains or advantages associated with adoption of tiered rates over simple metered rates. Further research is needed to account for the reasons. We would speculate that overall base price of water probably plays more of a factor than tiered rates themselves over the long run in dictating reductions in water use. We also would speculate that tiered rates employed year-round and not just seasonally would have a greater impact on use – but we would hasten to add that, with only two seasonal-rate districts in our study, we do not have enough data to draw a firm conclusion about this.

Fourth, some outlier communities spark interesting observations, as alluded to previously. For example, Santa Ana, an economically low-income and multi-ethnic community, had the lowest 2005 per capita consumption – not surprising since this is a community with small homes on modest lots. However, after an initial drop, by 2010, it showed a slight increase in residential water use. This underscores the fewer degrees of water savings available to low-income, low consuming residents.

Another outlier – but in a different direction – was affluent Yorba Linda, whose water district witnessed an increase in consumption of some 13 percent over the period 2006–10, and where single-family homes experienced a slight increase in usage. In this case, the number of water accounts decreased, but per-account use increased. Moreover, there was an increase, we believe, in the number of commercial users being added to the mix.

Finally, we found that water consumption differences per capita from 2005 to 2010 were also affected by the recent economic and foreclosure crises. Some districts in the county had fewer single-family home accounts in 2010 than in 2005, because many homes were either foreclosed and/ or abandoned (even though overall population in the county increased during the same period). We would speculate that there has been a slight shift across the county to multi-family residential units, helping to depress per capita consumption. There is ample evidence that in economically depressed water districts and local communities, when forecasting the future, we can expect a rebound in the number of single-family accounts and, thus, water usage. That being the case, we can estimate that decreases in water consumption in these communities – for better or worse – are temporary perturbations which future trends should see alleviated. In conjecturing what additional efforts water utilities might contemplate in response to growing demand pressures and climatic variability, particularly in states such as California that are committed by policy initiative to aggressively seek significant reductions in urban water use, utilities face a serious quandary. In short, utilities will need to continue to foster further reductions in water use in order to lower overall costs for imported water from outside the Southern California region. At the same time, however, more revenues will likely be needed by these same utilities to make up for losses in water consumption leading to lower revenues because they face continued needs to fix and upgrade existing water infrastructure. This will be an ongoing challenge, and is underscored by the fact that water rates are becoming a politically contentious issue throughout Orange County and Southern California as utilities seek to persuade ratepayers of the need for higher tariffs, particularly in light of the economic problems facing parts of the region. In part this is due to these above-noted economic challenges and, in part, due to public skepticism regarding what rate increases are going toward, coupled with a "sour" economy (Eades, 2012: 3; Allen, 2012: A33).

In looking toward policy solutions, three emerging issues are important. First, the precise urban form which growing megacities take can have a profound influence on the water–environment footprint. For instance, sprawling horizontal urban development imposes numerous problems, including paving of streets and commercial districts, and diminishing of groundwater recharge (for example, Lauer, 2008; Kamieniecki and Below, 2009).

Second, while a greater concentration of people in urban areas may lower per capita costs for many types of water infrastructure, the need to expand water supply and treatment networks over vast distances may increase the likelihood of unaccounted water loss (for example, Satterthwaite, 2000). Third, because urban supplies are often imported from outlying areas, pressures may be imposed on these adjacent regions (for example, Gandy, 2008; Tortajada and Casteian, 2003; Yusuf, 2007; Zérah, 2008).

And third, many megacities have learned – in sometimes impromptu and improvisational ways – how to mitigate the pressures of growth and climate variability on water stress. Innovations include better managing water demands, coupled with compensatory efforts for low-income groups who can least afford these innovations. Developed-country cities as diverse as Los Angeles, New York, Tokyo, and Melbourne all illustrate these challenges. While not a megacity, the latter's environmental footprint upon its surrounding region – particularly in light of Australia's recent Millennium Drought – are comparable to that of a megacity. These problems have also been observed in developing-country megacities, including Mexico City, Mumbai, Beijing, and Lagos, as well as in smaller, water-challenged urban centers such as Accra, Ghana; Alexandria, Egypt; and Bogota, Columbia (Adekalu et al., 2002; Gandy, 2008; Tortajada and Casteian, 2003; Yusuf, 2007; Zérah, 2008; Hua, 2012; Sule, 2003; Howe et al., 2011).

These experiences, moreover, illustrate another important point: every city can be located along a water-sustainability spectrum – a sort of transition span toward achieving integrated water management capacity. Evolution along this spectrum (Figure 6.5) is dependent upon several factors. The most important of these are: (1) a city's level of economic development; (2) the willingness of decision-makers to move from "path-dependent"

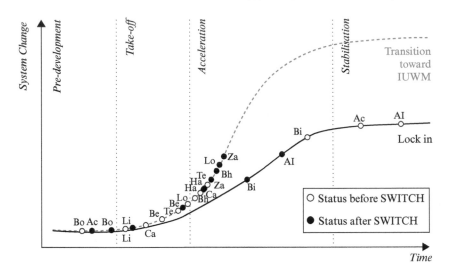

Note: Ac = Accra; Al = Alexandria; Be = Beijing; Bh = Belo Horizonte; Bi = Birmingham; Bo = Bogota; Ca = California; Ha = Hamburg; Li = Lima; Lo = Lodz; Te = Tel Aviv; Za = Zaragoza.

Source: Howe et al. (2011).

Figure 6.5   *Schematic of sustainable water management improves tomorrow's cities' health*

reliance on traditional big-engineering solutions to water problems; and (3) the willingness of decision-makers and others to consciously engage with regional partners and other stakeholders. In Los Angeles, New York, and Melbourne, for example, considerable attention has been paid to efforts to conserve water by working with regional partners in an effort to reduce reliance upon imports.

These lessons can be applied to many developing nation cities such as Mumbai, Beijing, Mexico City, Taipei, and others, as we will discuss. In our era, the need for such collaboration is made more problematic by global climate change and population growth, especially in developing nations that lack capacity to improve their water infrastructure (Vorosmarty et al., 2000; Jenerette et al., 2006; UN Human Settlements Programme, 2011; World Resources Institute, 2012). Collaboration can generate a collective, regional vision of how to manage water, and how to link scientists, users, and decision-makers.

NOTE

1. The Water Conservation Act of 2009, also known as California Senate Bill 7, requires water suppliers to decrease water consumption and improve water use efficiency among urban and agricultural users. Explicit goals were established (for example, urban per capita demand must be reduced 20 percent by 31 December 2020), while a mid-course reduction of 10 percent by 31 December 2015 is required. Some efficiency target variation is permitted depending on locational factors such as "weather, patterns of urban and suburban development, and past efforts to enhance water use efficiency" (California Senate Bill 7, 2009). The state's Department of Water Resources (DWR) is required to induce efficiency by establishing a framework to meet targets, measure efficiency, establish methods for "urban retail water suppliers to determine targets for achieving increased water use efficiency by the year 2020" and " . . . promote urban water conservation standards, . . . adopt best management practices, establish standards that recognize and provide credit to water suppliers that made substantial capital investments in urban water conservation since the drought of the early 1990s, and recognize and account for the investment of urban retail water suppliers in providing recycled water for beneficial causes" (California Department of Water Resources, Water Conservation Act of 2009, accessed 8 June 2013 at http://www.water.ca.gov/wateruseefficiency/sb7/. California Senate Bill No. 7, Chapter 4, November 2009).

PART III

The Path Forward: Technology, Infrastructure, Institutions, Practices

This final section discusses several ways forward in better managing water in cities across the world. Our focus is upon dual concerns: technology and policy. Chapter 7 examines the urban stream syndrome and how it can be alleviated, an especially important topic in the contemporary city that furnishes ways of managing both ecological and societal needs for water in beneficial ways.

Chapter 8 examines low-impact development approaches for both indoor and outdoor water uses. It also considers infrastructural innovations that sustain these low-impact developments and that are, in themselves, low carbon approaches. Chapter 9 examines the future of water governance and management for achieving a water-sensitive city. We consider new forms of governance for water management in cities and their surrounding regions with respect to water and, especially, polycentric forms of governance that encourage multiple jurisdictional collaborations for water management. Finally, Chapter 10 concludes our analysis with a discussion of future research needs of decision-makers.

7. Opportunities to satisfy urban water needs while addressing the urban stream syndrome[*]

Urbanization affects water in a number of ways, as we have seen. A little-explored area is associated with a reduction in stream health: a condition known as the urban stream syndrome (Figure 7.1) (Walsh et al., 2005b; Meyer et al., 2005; Wenger et al., 2009). Notable symptoms of this stream syndrome include altered stream flow, morphology, water quality, and ecosystem structure and function (Figure 7.1). While underlying causes of the urban stream syndrome will vary among catchments, its hydrologic symptoms are generally associated with: (1) replacing grassland and/or forests with impervious surfaces such as roads, parking lots, roofs, and sidewalks; (2) building drainage and flood control infrastructure to rapidly convey stormwater runoff to streams (so-called formal drainage systems); and (3) altering catchment water budgets (for example, through water imports and exports) (Figure 7.1) (Burns et al., 2012; Townsend-Small et al., 2013; Groffman et al., 2005; Hughes et al., 2014).

Increasing catchment imperviousness generally reduces infiltration and evapotranspiration of rainfall, while formal drainages increase the hydraulic connectivity between watersheds and streams (Gordon et al., 2013; Liu et al., 2008; Walsh and Kunapo, 2009). These two modifications have opposing effects on stream flow during wet and dry weather. During wet weather, the volume of stormwater delivered to a stream increases, the lag time between rainfall and storm flow gets shorter, and peak flow rate increases (Rose and Peters, 2001; Miller et al., 2014). During dry weather, stream flow decreases due to reduced infiltration over inter-annual time-scales (Hamel et al., 2013; Walsh et al., 2012), although there are exceptions. Water imports by way of diversions and transfers can increase dry weather stream flow by increasing: (1) perennial discharge of wastewater effluent and nuisance runoff; and/or (2) groundwater seepage from leaks in subterranean drinking water supply and sewage collection pipelines. Management of surface water impoundments (for example, dams and reservoirs) can also increase dry weather stream flow (Hopkins et al., 2015). All of these watershed modifications, in addition to altering stream

A. Symptoms

flow   morphology   water quality   ecology

B. Hydrologic Drivers

imperviousness   formal drainage   stream modification   imported water

C. Hydrologic Remedies

| unlined biofilter | permeable pavement | green roof | rain tank |

Notes: (A) Symptoms include: (1) altered stream flow (base flow, peak flow, annual runoff volume, flow variability); (2) altered stream morphology (stream width, depth, complexity, and disconnection from the riparian zone, hyporheic zone and flood plain); (3) impaired water and sediment quality (trash, nutrients, dissolved oxygen, toxicants, suspended solids, temperature); and (4) shifts in biological composition (loss of native species, reduction in sensitive species, increase in tolerant species, increase in invasive species) and loss of ecosystem services (organic matter retention and processing, nutrient removal, primary production, and respiration). (B) Causes include: (1) replacing grassland and/or forests with impervious surfaces such as roads, parking lots, roofs and sidewalks; (2) building stormwater drainage and flood control infrastructure to rapidly convey stormwater runoff to streams (formal drainage systems); (3) reducing stream complexity by burying, straightening and concrete-lining streams; and (4) altering overall water and sediment budgets through water importation, the construction of debris dams, and surface water impoundments. (C) Examples of LID technologies that can potentially address the hydrological challenges associated with the urban stream syndrome include unlined technologies that infiltrate stormwater runoff (e.g., unlined biofilters and permeable pavement) and technologies that harvest and export stormwater runoff from the catchment (e.g., green roofs and rainwater tanks used for irrigation or indoor toilet flushing). Top row includes images of urban creeks and drains in Orange County, California (from left to right: San Diego Creek, Costa Mesa Channel, Fullerton Creek, and a drain in the City of Irvine). Middle row includes two streetscapes and a buried stream in Orange County California, and Parker Dam at the start of the Colorado Aqueduct on the California–Nevada border. Bottom row includes an unlined biofilter n Melbourne (Australia); permeable pavement in Westminster, California; green roof on a public building in Houston, Texas; and a rainwater tank in Melbourne (Australia).

Source: After Askarizadeh et al. (2015) with permission.

Figure 7.1   Symptoms, causes and cures of hydrologic perturbations associated with the urban stream syndrome

hydrology, degrade stream water quality by raising stream temperature, changing the balance of nutrients, carbon, and oxygen in a stream, and facilitating the mobilization and transport of fine sediments, chemical pollutants, and human pathogens and their indicators (Miller et al., 2010; Reeves et al., 2004; Grant et al., 2011; Rippy et al., 2014; Surbeck et al., 2010). Changes in water quality and hydrology (both symptoms of urbanization) act in concert to affect stream morphology, stability, ecology, and chemistry (Ometo et al., 2000; Morse et al., 2003).

Watershed urbanization is commonly quantified using two measures: total imperviousness and effective imperviousness (Welty et al., 2009; Booth and Jackson, 1997). Total imperviousness is the fraction of catchment area covered with constructed impervious surfaces such as asphalt and roofs. Effective imperviousness represents the impervious fraction of the catchment area with hydraulic connection to a stream through a formal drainage system. The latter is generally a better predictor of stream water quality, ecological health, and channel form (Hatt et al., 2004; Taylor et al., 2004; Vietz et al., 2014). Total imperviousness does not take into account whether flow from an impervious surface is conveyed directly to a stream, or instead drains to adjacent pervious areas where opportunities for filtration, infiltration, and flow attenuation are provided. The ecological condition of streams typically exhibits a wedge-shaped dependence on total imperviousness: streams in catchments with low total imperviousness exhibit a range of ecological conditions (from degraded to healthy) that narrows with increasing total imperviousness due to reduction in the maximum attainable stream health. Effective imperviousness exhibits a less variable negative correlation with stream ecological condition, water quality and channel form (Walsh and Kunapo, 2009).

The negative correlation between effective imperviousness and stream health raises the question: can hydrologic symptoms of the urban stream syndrome be prevented or reversed through urban forms that keep effective imperviousness low? Effective imperviousness can be kept low as an urban community develops (or reduced through retrofits of an already developed catchment) using technologies that intercept runoff from impervious surfaces at a variety of scales (Fletcher et al., 2008; Fletcher et al., 2013a).

The intercepted runoff can be infiltrated to support groundwater (for example, with unlined biofilters and permeable pavement), exported to the atmosphere by evapotranspiration (for example, using green roofs, rain gardens, vegetated swales, wetlands, and urban forests), redirected from storm sewer systems to pervious surfaces (for example, with downspout disconnection systems), and/or exported through the sanitary sewer system to downstream receiving waters (for example, using rainwater tanks for toilet flushing) (Figure 7.1). These environmentally sensitive stormwater management systems go by a variety of names, as we noted in Chapter 1. In this chapter, we have chosen to use the term 'low-impact development' or LID technologies – although as we will see, they are increasingly being labeled "green infrastructure."

Acquiring and maintaining public support for LID technologies requires demonstrating that they are effective at minimizing flood risk and the negative impacts of urbanization on human and ecosystem health (Poff and Zimmerman, 2010). We now offer: (1) a conceptual framework for supporting LID selection and evaluation; (2) technologies available for stormwater infiltration and harvesting; and (3) implementation challenges including maintenance, climate change, path dependence, water quality, human health, and site-specific constraints.

WATERSHED-SCALE URBAN WATER BALANCE

In many countries, stormwater regulations place limits on the peak flow rate or high flow duration allowed to enter a stream from individual properties (Booth and Jackson, 1997). To comply with these regulations, property owners typically install stormwater detention ponds that capture and slowly release runoff from large storms (Guo, 2001). There are a number of well-documented problems with this approach, including: (1) the simultaneous release of stormwater from many properties within the catchment can cause downstream peak flows to exceed pre-development conditions and erode downstream channels, even if the peak flows from individual properties remain within regulatory limits; (2) reduced infiltration associated with impervious surfaces cuts off the primary means by which water is normally supplied to a stream (through subsurface flow paths and resupply of shallow groundwater), and detention ponds do not typically address the problem; and (3) the superposition of post-storm flows from multiple detention basins in a catchment distorts downstream dry weather flow regimes (Petrucci et al., 2013; Petrucci et al., 2014; Emerson et al., 2005). While a number of stream "sustainability" measures have been proposed, controlling (and ideally eliminating) the volume of stormwater runoff flowing to a stream through formal drainage systems is a prerequisite for maintaining and restoring the pre-urban flow regime (discussed below) (Reichold et al., 2010; Giacomoni et al., 2012).

URBANIZATION IMPACTS ON WATERSHED-SCALE WATER BUDGETS

Over annual timescales, subsurface flow constitutes the majority of stream flow in most natural watersheds, including during storm events (Bhaskar et al., 2015). In this context, subsurface flow (sometimes referred to as "old water") is defined as rainfall that infiltrates and flows to a stream through shallow groundwater. By contrast, the contribution of overland flow to annual stream flow is generally small in natural catchments (Burns et al., 2013; Booth, 1991; Buttle, 1994).

Urbanization perturbs this situation in a number of ways (Figure 7.2): (1) by generally decreasing evapotranspiration (ET) by replacing forests and/or grassland with impervious surfaces (one exception to this rule is

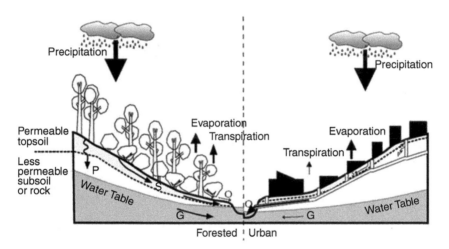

Notes: In the pre-urban state (left) a large fraction of annual precipitation is returned to the atmosphere as evaporation and transpiration (evapotranspiration, ET). The relatively small portion of precipitation that is not returned to the atmosphere flows to receiving waters (rivers, lakes, estuaries, coastal ocean) through shallow and deep groundwater flow. By contrast, in the urbanized environment ET is reduced (because grasses and forests have been replaced with impermeable surfaces) and therefore more water is left on the ground compared to the pre-urban state. Further, the impervious surface reduces the fraction of precipitation that infiltrates into shallow and deep groundwater. The net result of these two processes is reduced baseflow during dry weather (because the re-supply of groundwater has been reduced) and a massive increase in runoff during storm events. Rather than correct the root cause of these two perturbations, cities respond by building formal drainage systems which rapidly convey extremely large volumes of poor-quality stormwater to receiving waters. Thus, hydrological alterations directly associated with the built environment are a major cause of receiving water quality problems and the urban stream syndrome.

Figure 7.2   Changes to catchment hydrology associated with urbanization

regions where significant water importation occurs, in which urbanization can increase ET above historical levels) and increasing the flow of water to streams and receiving waters (because the water that used to return to the atmosphere as ET now goes directly to the receiving water); and (2) by altering how water is delivered to the stream, from subsurface flow paths in the pre-urban state to a mixture of subsurface flow and overland flow from effective imperviousness in the urban state.

Maintaining Pre-Urban Hydrology through Infiltration and Harvesting

Two categories of LID technologies can be deployed to support pre-urban stream flow as a catchment develops. The first type, infiltration-based LID technologies, transfer stormwater runoff to the subsurface where it can recharge groundwater supplies and provide base flow for local streams. The second type, harvest-based LID technologies, capture the remaining runoff (that is, the stormwater not infiltrated) and use it for purposes that keeps it out of the stream (for example, irrigation of ornamental plants and toilet flushing) (Grant et al., 2012). The challenge is deploying the right number and mix of these two LID types within cities; namely, enough infiltration- and harvest-based LID technologies to compensate for the infiltration and evapotranspiration lost when forests and grasslands were replaced with impervious surfaces. Another benefit of deploying both types is that the hydrology of the local stream is unchanged as the catchment urbanizes because: (1) subsurface flow to the stream is maintained at pre-development levels, and (2) no stormwater runoff flows overland to the stream via effective imperviousness.

Tailoring Infiltration and Harvesting to Specific Regions

The Walsh bucket model suggests that LID technologies have the potential to remedy hydrologic symptoms associated with the urban stream syndrome. An interesting consequence of the Walsh bucket model is that the relative proportion of runoff volume that should be infiltrated and harvested is constant and depends on only two variables: the mean annual rainfall and the fraction of the pre-urban catchment area covered with forest. The net result (see Askarizadeh et al., 2015 for details) is that for most climate and pre-urban conditions, considerably more stormwater should be harvested than infiltrated. Thus, if we were to look for an urban form that saves potable water (for example, through potable substitution) and simultaneously protects surrounding aquatic ecosystems, the emphasis should be on LID technologies that harvest stormwater over a wide range of climates (biofilter configuration A in Figure 7.3).

LID Technologies for Maintaining or Restoring Pre-Urban Hydrology

Translating theory to practice will require a diverse set of LID technologies tailored to: (1) capture all stormwater runoff before it enters the stream; and (2) infiltrate and/or harvest the captured runoff in the proper proportions. In practice, many different factors go into the selection of LID technologies (for example, flood protection, operation and maintenance costs, site-specific constraints, and human and ecosystem co-benefits) (Facility for Advancing Water Biofiltration, 2009).

We contend that the first-order concern in LID technology selection should be maintaining (or restoring) pre-urban flow regimes, with secondary consideration given to other constraints and benefits. Accordingly, in this subsection we classify several popular LID technologies relative to the three end points that underpin the Walsh bucket model presented above: the percentage of runoff volume harvested, infiltrated, or left as overland flow (represented by vertices of the ternary diagram in Figure 7.4).

Examples of infiltrative systems include infiltration trenches and permeable pavement (represented in Figure 7.4 by a blue arrow, light blue arrow, light blue dashed box and brown arrow). Infiltration trenches and permeable pavement without under-drains (that is, drains that collect some fraction of the outflow from a system) infiltrate the highest percentage of runoff (60–100 percent runoff removed) (Charlesworth et al., 2003; Brattebo et al., 2003; Shuster et al., 2005).

Permeable pavement with under-drains infiltrate less runoff because a fraction of outflow is piped to the storm sewer system (25–66 percent runoff removed; light blue arrow in Figure 7.4). Re-routing this piped fraction to a storage facility can transform permeable pavement with under-drains from infiltration to hybrid systems (that is, technologies that both infiltrate and harvest; light blue dashed box, in Figure 7.4), assuming that the captured water is used for irrigation (evapotranspiration) or in-house activities (for example, toilet flushing) that transfer the water to the sanitary sewer system. While treated stormwater is rarely used for domestic purposes in the U.S., such systems are actively being trialed in southeast Australia (Low et al., 2015).

Harvesting Technologies

Examples of harvest-based LID include green roofs, rainwater tanks and wetlands (shown as a pink arrow, green arrow, and orange dashed arrow,

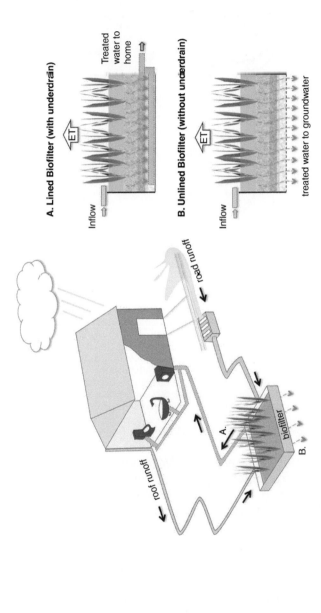

A. Lined Biofilter (with underdrain)

Treated water to home

ET

Inflow

B. Unlined Biofilter (without underdrain)

ET

Inflow

treated water to groundwater

roof runoff

road runoff

A.

B.

biofilter

Notes: In the example illustrated here a biofilter is configured to receive both roof and road runoff. In a harvest configuration, treated water from the biofilter can provide non-potable water to the home for toilet flushing, laundry and hot water supply (lined biofilter with underdrain, A). In an infiltration configuration, the biofilter supports groundwater recharge and stream baseflow (unlined biofilter without underdrain, B). In both configurations, a portion of the water processed by the biofilter is lost to the atmosphere through evapotranspiration (ET), another form of harvesting. Colored layers in the biofilters (upper and lower right panels) delineate ponding zone (blue), filter media (brown), transition layer (light brown), and gravel layer (gray).

Source: Used with permission from Azkarizadeh et al. (2015).

Figure 7.3   Biofilters are a hybrid LID technology that can be tuned to achieve different levels of stormwater harvest and infiltration

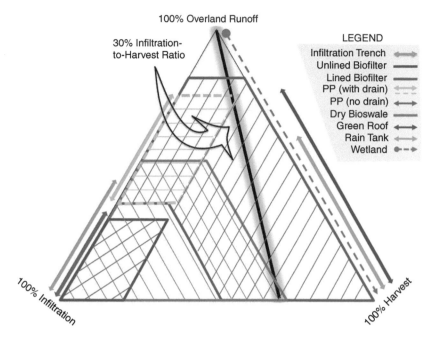

Notes:
The abbreviation "PP" refers to permeable pavement. The designation "with drain" refers
to systems in which treated effluent can be routed to storage facilities for non-potable uses,
such as garden irrigation and toilet flushing. The designation "without drain" refers to
systems in which treated effluent leaches directly into the subsurface. Arrows along the side
of the ternary diagram denote systems that are used primarily for infiltration (left leg of the
triangle) or for harvesting (right leg of the triangle). Polygons indicate hybrid systems that
can be "tuned" to provide specific infiltration-to-harvest ratios. Solid colored lines reflect
observed performance, whereas colored dashed lines denote theoretical performance (i.e.,
the performance is possible but not documented). The thick black line with a blue halo
marks the location of hybrid systems that achieve a 30% infiltration-to-harvest ratio.

Source: Adapted with permission from Askarizadeh et al. (2015).

Figure 7.4   *A ternary representation of field and laboratory data on
the performance of popular LID technologies relative to
percentage of runoff volume infiltrated (lower left vertex),
harvested (lower right vertex), and allowed to flow to the
stream through connected imperviousness (top vertex)*

respectively, in Figure 7.4). A broad range of harvest efficiencies have been
noted for green roofs (23–100 percent runoff removed) (Berndtsson, 2010;
Nicholson et al., 2009; Coombes and Kuczera, 2003; Kahinda et al., 2007;
Persson et al., 1999; Rousseau et al., 2008; Ahiablame et al., 2012). Green

roofs export runoff mostly in the form of evapotranspiration, with the soil/ media matrix dominating export in the winter (low harvest: ~ 34 percent runoff removed) and the "green" component contributing to export in the summer (high harvest: ~ 67 percent runoff removed).

Rainwater tanks harvest between 35 and 90 percent of runoff on average depending on the ratio of tank size to roof area, storm frequency and duration, the number of acceptable rainwater uses (for example, toilet flushing, clothes washing, hot water supply, or garden irrigation), and building occupancy. Human use of rainwater is expected to be higher in multi-story residential and office buildings than in commercial/industrial buildings, given the greater number of inhabitants per unit area of imperviousness. Although wetlands typically export relatively small volumes of runoff in the form of evapotranspiration (0–3 percent runoff removed), outflow can be tapped for human use, substantially increasing the overall percentage of runoff harvested. Upwards of 50–100 percent harvest has been reported for wetland systems in South Australia and New South Wales, resulting in potable water savings of $120 000 to $663 120 year $^{-1}$ (in 2006 AUD) (Hatt et al., 2006).

Hybrid Technologies

LID technologies that both harvest and infiltrate stormwater runoff, or "hybrid technologies," appear as polygons in Figure 7.4. Examples of hybrid technologies include unlined biofilters (no underdrain, blue polygon), partially or completely lined biofilters (with underdrain, red polygon), and dry bioswales (unlined with an underdrain, green polygon). The term "dry bioswales" refers to swales that are intended to dry out between storms. Two configurations for a household biofilter (lined with an underdrain versus unlined with no underdrain) are illustrated in Figure 7.3.

Few studies have quantified the percentage runoff harvested through evapotranspiration for hybrid systems. Values as low as 2–3 percent runoff removal have been reported for unlined biofilters; however, these percentages may be low because a substantial portion of infiltrated runoff passes into upper soil layers where additional (unquantified) evapotranspiration may occur. Higher evapotranspiration values (> 19 percent runoff removed) have been reported in lined biofilters (Hamel et al., 2011). Thus a tentative range for percentage runoff harvested via evapotranspiration across biofilters (lined and unlined) is 2–19 percent. Unlined biofilters are primarily infiltration systems, with evapotranspiration constituting their primary contribution to harvest (total runoff removed ranging from 73 to 99 percent; evapotranspiration: 2–19 percent; and infiltration:

71–97 percent). In contrast, lined biofilters are often used to treat storm-water prior to discharge to a storm sewer system; the treated effluent can also be captured and stored for subsequent human use, increasing harvest potential (total runoff removed ranging from 20 to 100 percent; evapotranspiration: 2–19 percent; human use: 0–80 percent; and infiltration: 1–63 percent). Dry bioswales are effective for harvesting and infiltrating runoff, with near 100 percent runoff removal achieved over a broad combination of infiltration and harvesting percentages (total runoff removed ranging from 46 to 100 percent; evapotranspiration: 2–19 percent; human use: 0–54 percent; and infiltration: 27–96 percent). The effectiveness of dry bioswales for harvesting runoff can be attributed to their relatively large surface area to catchment area ratio, compared to other hybrid systems (Hirschman et al., 2008; Hamel et al., 2011; Li et al., 2009).

Matching LID Technologies to Stormwater Management Goals

In many locales, the emphasis should be on harvest-based LID technologies. This may prove challenging in practice, because distributed harvest systems that capture stormwater runoff at its source (for example, rainwater tanks and green roofs) only treat one form of impervious area (rooftops), leaving untreated the runoff from other, potentially much more extensive imperviousness (for example, roads, parking lots, and residential driveways). While regional (or end-of-catchment) LID such as wetlands can be employed to harvest the remainder, this approach is at the expense of water quality in reaches upstream of regional facilities. Alternatively, runoff from roads and driveways can be captured and harvested using distributed hybrid systems (for example, lined biofilters, dry bioswales, and permeable pavement with underdrains) configured to provide non-potable water for human use (configuration A in Figure 7.3).

At the parcel scale, LID technologies (or combinations of LID technologies) can be selected to match catchment-scale goals for the volume of runoff to be infiltrated and harvested. For example, consider a case where the target infiltration-to-harvest ratio (which best mimics pre-urban hydrology) is 30 percent. That particular infiltration-to-harvest ratio translates to a straight line in Figure 7.4 (see thick black line). In practice, this infiltration-to-harvest ratio can be achieved by selecting hybrid technologies that cross or enclose the line (for example, lined biofilters "tuned" to achieve the 30 percent target) and/or by a combination of infiltration and harvest technologies designed to operate toward the harvesting end of the spectrum (for example, treatment trains consisting of large rain tanks that overflow to unlined biofilters) (Burns et al., 2012).

A prime outcome of the watershed water balance described above is that, for most areas of the world, restoring this balance will require a focus on harvest-based LID technologies. A win–win example is using harvested rainwater and road runoff for in-home activities (for example, for toilet flushing, laundry, and hot water supply, configuration A, Figure 7.3), thereby protecting streams and reducing potable water consumption (Grant et al., 2012). However, in the U.S. a number of institutional barriers presently limit the indoor use of non-potable water. These include: (1) low uniform water prices that create an environment where consumers and developers have little incentive to invest in schemes to reduce potable water consumption, although this is changing in the southwestern U.S.; (2) plumbing codes that do not explicitly address rainwater use or inadvertently prohibit it by requiring that downspouts be connected to the storm sewer collection system; (3) a patchwork of local, state, and federal regulations with various and conflicting treatment standards; (4) prohibitions against indoor use of non-potable water in some locales that prevent local water utilities from sponsoring such schemes; (5) different interpretations of who owns stormwater runoff, with some states (for example, Colorado) prohibiting residential capture and reuse of stormwater on the premise that all rainfall has been already allocated to downstream users; and (6) resistance from drinking water providers over concerns that widescale adoption of rainwater and stormwater harvesting may endanger public health, or lead to revenue loss (Roy et al., 2008; Kloss, 2008; Garrison et al., 2011).

While public health concerns are often cited as a barrier to the adoption of harvested rainwater and stormwater for non-potable uses in the U.S., the scientific evidence (and practical experience) generally does not support that contention. Public health concerns stem from the fact that both sources of water can harbor micro-organisms that cause human disease. Human infection depends on multiple factors – pathogen type and load, the mode of exposure, and susceptibility – that are best assessed through epidemiological studies and/or a quantitative microbial risk assessment (QMRA) framework that includes hazard identification, exposure assessment, dose-response assessment, and risk characterization (Ahmed et al., 2011; Grebel et al., 2013; Lim and Jiang, 2013; Lim et al., 2015).

An epidemiological study of children in rural South Australia found that drinking roof harvested rainwater posed no more risk of gastroenteritis than drinking water from a reticulated supply. However, concerns have been raised about the study's sensitivity (ability to detect an effect against background rates of infection) given that only 1016 people participated. QMRA studies, which have been advocated as a more sensitive alternative to epidemiological investigations, indicate that minimally treated stormwater and rainwater may be acceptable for certain in-home uses, such as toilet flushing (Hayworth et al., 2006).

Rainwater also appears acceptable for garden irrigation and showering. However, the suitability of stormwater for these purposes is less well resolved. Across the board, proper design and maintenance of collection systems as well as appropriate disinfection measures such as UV disinfection and chlorination are necessary to achieve public health targets for in-home use. Currently, more than two million Australians use roof-harvested rainwater for potable or non-potable supply, and the State of Victoria now requires new homes to have a rainwater tank for garden watering and in-home uses such as toilet flushing (although solar hot water heating can be installed as an alternative, suggesting that this instrument has a broad focus on "sustainability," rather than a specific focus on water management). Australia's ongoing experiment with rainwater tanks (and more recently biofilters) should provide a wealth of data and experience with which health officials around the world can objectively evaluate the risks and benefits for in-home use.

Site-specific constraints may also impede infiltration schemes. For example, the City of Irvine (California) discourages stormwater infiltration at certain locations due to low soil permeability, locally perched shallow groundwater, and concern that groundwater contaminants (such as selenium) may be mobilized into local streams or the deep aquifer used for potable supply (Stephens and Associates, 2013). This concern is shared by the Orange County Water District (which manages the local groundwater basin that supplies drinking water to more than two million residents) and the Orange County Healthcare Agency (which manages public health for the county), and is enshrined in county regulatory statutes. Thus, for this particular region of Southern California, infiltration may be feasible in only a few locations and under fairly strict control; for example, at large centralized facilities strategically placed to facilitate runoff treatment and recharge to deep groundwater aquifers (Reilly et al., 1999).

Evaluating the Efficacy of Green Infrastructure

Once LID technologies have been selected and implemented, monitoring programs are needed to ensure their goals are being met. A number of recent reviews summarize evaluation approaches for gauging the effects of land-use and land-cover change and LID interventions, watershed-scale hydrologic budgets and stream flow (O'Driscoll et al., 2010; Machiwal and Jha, 2009). The gold standard for assessing the hydrologic impact of land-use change is paired (or triplicated) catchment studies, in which the catchment of interest is paired with a control catchment (and a reference catchment, in the case of a triplicate design) of similar climate and physiography (Brown et al., 2005; Watson et al, 2001). There is a long history of using paired catchment studies to assess the impact of vegetation change on catchment hydrology, but the technique has been applied only recently to assess the impacts of LID interventions on stream health (Zhang et al., 2001; Zhang et al., 1999).

Such studies demonstrate that adopting LID technologies for storm-water management (over conventional centralized retention and detention basins) markedly improves the hydraulic performance of streams, as measured by higher baseflow, lower peak discharge and runoff volumes during moderate storms, increased lag times, and retention of smaller, more frequent precipitation events. These field results are generally supported by modeling studies, although centralized stormwater control measures may perform better than distributed LID systems for controlling peak discharge from large storms, a problem that could presumably be overcome by proper LID technology placement and design (Hood et al., 2007; Selbig and Bannerman, 2008; Bedan and Clausen, 2009).

Not surprisingly, none of the urban stormwater management approaches perform as well as un-urbanized (reference) catchments. Thus, it can be argued that the best approach for protecting stream health is to place strict limits on urban development within a catchment. Short of this goal, however, distributed LID technologies should be used for managing stormwater runoff (Loperfido et al., 2014).

The next frontier is paired catchment studies that evaluate how LID interventions simultaneously influence the hydrologic, water quality, and ecological response of streams. One example is Little Stringybark Creek in Melbourne, Australia. In collaboration with a local water utility, researchers developed a financial incentive scheme to encourage homeowners to install rainwater tanks and unlined biofilters, and worked with the local municipality to install larger neighborhood-scale infiltration and harvesting systems. To determine if these retrofits are impacting flow, water quality, and ecology in Little Stringybark Creek, researchers are employing a "before/after control reference impact" (BACRI) study, consisting of the study catchment (where LID technologies are implemented), two urban control catchments (with similar levels of effective imperviousness, but where LID technologies are not implemented), and two non-urbanized reference catchments representing natural conditions. While such experiments are ambitious and challenging, they are a rigorous field test for how well LID technologies insulate streams from catchment urbanization. The project has already generated important lessons in relation to community engagement, institutional aspects, and the performance of LID technologies in flood reduction. There are some early signs that the retrofit may be improving water quality in the creek (Walsh et al., 2005b; Bos and Brown, 2015; Burns et al., 2015).

CONCLUSIONS: PATH-DEPENDENCE AND THE URBAN STREAM SYNDROME

Many factors may make the urban stream syndrome context- and path-dependent. By this we mean that the state of a stream depends not only on the extent of LID intervention (as measured, for example, by the volume of stormwater harvested and infiltrated) but on the environmental context and historical path by which the watershed arrived at its current state. One form of path dependence is called "cognitive lock-in," wherein positive feedback between the societal perception, management, and the physical and biological condition of a stream varies within communities depending on their phase of economic development (Figure 7.5). The term "cognitive lock-in" originates from the field of social psychology, where it has been applied to understanding consumer habits and choices with respect to a product or service. The idea is that repeated consumption or use of a product results in a (cognitive) switching cost that increases the probability that a consumer will continue to choose that product or service over alternatives (Ferguson et al., 2013; Johnson et al., 2003; Murray and Häubl, 2007).

As applied here, cognitive lock-in can affect stream health in positive or negative ways (Figure 7.5). If a community perceives their stream is a threat (for example, due to the damage it might cause by flooding), local managers may be pressured to enact policies that degrade a stream's aesthetic and eco-logical value (for example, through installation of formal drainage with high effective imperviousness, and stream burial), unintentionally reinforcing neg-ative perceptions of the stream as a drain (red loop in the figure). Conversely, if a stream is perceived as a valuable asset, local managers may respond by enacting policies that protect the stream from urbanization, reinforcing posi-tive perceptions of the stream as an asset through increased property value and the provision of green space and other ecosystem services (green loop).

Examples of cognitive lock-in abound in stormwater management, and its manifestations are evident in urban centers as diverse as Los Angeles, Paris, Moscow and Melbourne. A common pattern is that, as cities indus-trialize, prevailing public values call for harnessing and restraint of urban rivers for flood control and property development (favoring the red loop), while post-industrial development leads to demand for restoration of rec-reational, aesthetic, cultural heritage, and ecological values (favoring the green loop).

Path dependence can also play a role in urbanization thresholds. These thresholds are defined as critical levels of urban intensity and they are measured by effective imperviousness, road density, or the metropolitan area national urban intensity index, MA-NUII. Exceeding these thresholds causes symptoms of the urban stream syndrome to begin as the watershed

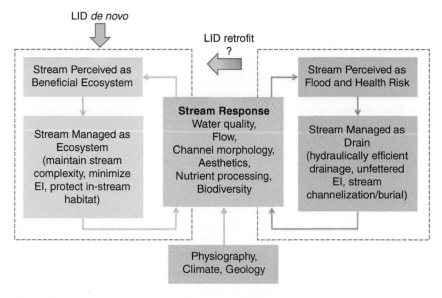

Note: The left loop may be more likely to occur if LID technologies are incorporated into an urban space as a city develops ("LID de novo"). Retrofitting an already developed area with LID technologies may or may not trigger a transition from the right loop to the left loop ("LID retrofit").

Source: Adapted with permission from Askarizadeh et al. (2015).

Figure 7.5 *Social-ecological feedback loops can lead to "cognitive lock-in" in which streams are maintained in either a degraded state (because they are perceived primarily as storm drains, right loop) or healthy state (because they are perceived as ecologically valuable assets, left loop)*

urbanizes, or, conversely, declines as the community adopts and retrofits with LID technologies (Roy et al., 2014). Evidence for urbanization thresholds comes from comparing metrics of stream health (hydrology, water quality, and/or ecology) across two or more nearby catchments with different levels of imperviousness (that is, paired catchment studies). For example, Walsh and colleagues (2012) found that stream health (as measured by hydrologic indicators, water quality, and biodiversity) was good in two catchments with low effective imperviousness (< 1 percent), but poor in two nearby catchments with elevated effective imperviousness (5 and 22 percent). Effective imperviousness thresholds of up to10 percent have been associated with significant degradation in one or more stream metrics.

As noted by Hopkins and colleagues (2015), this threshold may reflect the tendency of urban communities to transition from mostly informal (unsewered) drainages below 10 percent to mostly formal (sewered) drainages above 10 percent imperviousness (although their measure of imperviousness is a satellite product that may not equate to effective imperviousness). Collectively, such studies suggest that preventing the urban stream syndrome requires keeping effective imperviousness well below 10 percent and perhaps below 1 percent, although there is considerable study-to-study variability depending on climate, physiography, geology, land-use, and stream history (Jacobson, 2011; Roy et al., 2014; Utz et al., 2011; Cuffney et al., 2010).

Threshold variability is an inevitable consequence of the environmental context- and path-dependent nature of stream health. In some streams there may be no urbanization threshold below which the effects of urbanization cannot be observed. As part of the U.S. Geological Survey's National Water-Quality Assessment (NAWQA) Program, Cuffney and associates evaluated the impact of urbanization on in-stream invertebrate assemblages (a measure of stream ecosystem structure and function) across urban-to-rural gradients in nine metropolitan areas of the U.S. They found that invertebrate assemblages were strongly related to urban intensity (MA-NUII), but only when the urban development occurred within forests or grassland. A much weaker (or non-existent) correlation was observed in areas where agriculture or grazing predominated, presumably because those streams were already degraded. Importantly, in forests and grassland there was no urbanization threshold below which ecosystem assemblages were resistant to urbanization. Even low levels of imperviousness (5–10 percent) were associated with "significant assemblage degradation and were not protective" (Williams, 2001).

It is not surprising that imperviousness thresholds are not always present, since effective imperviousness is only one of many stressors that can negatively impact urban stream health. For example, salinization exerts an enormous ecological toll on streams worldwide. While road runoff clearly contributes to the problem (particularly in northern climates where salt is used for de-icing roads), there are other sources of salt that would not be eliminated by reducing effective imperviousness alone (for example, return flows from irrigated agriculture) (Kelly et al., 2008; AghaKouchak et al., 2014). Other examples of urban stream stressors include loss of riparian habitat and tree canopy, impoundments that alter flow regimes and elevate temperatures, point-source discharges of nutrients, heavy metals, and contaminants of emerging concern, to name a few. In short, reducing effective imperviousness may be a necessary, but not sufficient, condition for curing the urban stream syndrome in some catchments.

For all of the reasons stated above, it is difficult to predict the imperviousness threshold (if one exists) at which stream conditions will markedly improve as an urbanized catchment undergoes an LID retrofit. Shuster and Rhea (2013) reported a small but significant improvement in the hydrology of a small suburban creek (Shepherd Creek, Cincinnati, Ohio) after installing 165 rain barrels and 81 unlined biofilters in the 1.8 km² catchment. These measures reduced effective imperviousness by approximately 1 percent, mostly from roofs. However, a follow-up study by Utz et al. (2011) of the same LID retrofit reported little change in water quality and ecology of the stream compared to a control stream in the nearby catchment.

The authors suggest a number of possible explanations for the lack of a water quality and ecological response, most notably that, despite the relatively large investment in LID retrofits, effective imperviousness in the catchment was not reduced to levels where improvements in stream health would be expected (after retrofits, the effective imperviousness in the Shepherd Creek catchment was still above 10 percent). The authors concluded that, "additional research is needed to define the minimum effect threshold and restoration trajectory for retrofitting catchments to improve the health of stream ecosystems" (Utz et al., 2011). Ongoing retrofits in the Little Stringy Bark Creek project, which will reduce effective imperviousness below 1 percent, may eventually shed light on this important issue.

While LID technologies are not a cure-all for every symptom of the urban stream syndrome everywhere, they do address the critical hydrologic and geomorphic symptoms of the disease while providing myriad co-benefits and subsidiary ecosystem services, including water quality improvement, flood protection, green space, recreation and aesthetic value, wildlife habitat and corridors, carbon sequestration, pollination services, urban heat island cooling, and a much-needed supply of non-potable ("fit-for-purpose") water in drought-prone areas such as southeast Australia and the southwest U.S. (Endreny, 2008; Coutts et al., 2013).

NOTE

* This chapter is largely extracted from Askarizadeh et al. (2015).

8. Low-impact development: indoor and outdoor innovations

Innovations intended to conserve water or use less of it to meet the same needs are referred to as *demand-side approaches*. They include direct household metering of water, variable rate structures that charge more for more volumes of water used by consumers, and mandatory water-appliance retrofitting or other mandatory conservation measures. Demand-side approaches may burden economically disadvantaged groups by ignoring their ability to pay for water, or forcing them to install high-cost, lower water-using appliances. These approaches are among well-established "low-energy" approaches to urban water management. They are not – as we discuss – the only approaches.

RATE STRUCTURE AND RELATED INNOVATIONS

Metering measures water consumption at the point of consumer use. The introduction of metering in domestic uses where it has not been employed previously, such as multi-family apartment complexes, has in some locations led to water savings of as much as 20–40 percent compared with previous volumes. However, metering may penalize lower-income residents for the simple reason that the type of residential housing stock that lacks metering, and for which it is prescribed, are apartment buildings and other domiciles that disproportionately house lower-income groups. For people on fixed incomes, moreover, particularly if they are also charged for the cost of meter installation, costs can be burdensome.

One study of installation of water meters in Argentina found that public resistance in Salta province led to vandalism of newly installed meters, protests, and refusal to accept metering at neighborhood meetings. These were prompted by the fact that metering was introduced at the same time as aggressive enforcement of bill payment, "tiered" rates, and household charges for meter installation. A larger fear was that meters would not be accurately read and that residents would be charged for excess water usage when, in fact, the culprit was poorly maintained, leaking plumbing systems (Post, 2009).

Installation of meters in low-income housing areas may also penalize apartment owners, discouraging additional apartment housing investments that might benefit additional lower-income families. In effect, the installation of meters can sometimes *outweigh* water savings. Finally, there is a social cost of metering or otherwise encouraging the conservation of water. It may actually encourage undesired community growth by convincing communities that there is additional "saved" water that can now accommodate additional population (negating the savings). In some parts of the U.S., California being an example, state law mandates that new dwellings have water meters. However, utilities are not required to read them, or to bill at metered rates (Hanak et al., 2012b).

Increasing block rate (IBR) pricing charges customers more per unit of water used once their volume of use exceeds an average derived use level (that is, a "conservation base"). In principle, IBR pricing assumes that the greater one's income, the greater one's water use – valid when applied to homeowners who practice widespread outdoor uses (for example, landscaping, pools). Thus, it would, on the surface, appear to be equitable as well as efficient. Where introduced, water savings of somewhere between 10 and 15 percent have been reported. However, like metering, IBR may not account for ability to pay, especially for those on fixed incomes who, for health or other reasons, use more water. Moreover, it may create a political conundrum for water utility districts whose elected boards prefer voluntary as opposed to mandatory measures to conserve water.

In practice, IBR is not always practiced in areas that need it. In California, for example, two-thirds of the population of the state's southern coastal area pay increasing block rates, and other incentives for conservation are fostered by utilities such as the Metropolitan Water District of Southern California, which has spent more than \$185 million over the last decade encouraging water customers to install water-efficient appliances, plant drought-resistant landscapes, and reduce overall water use. In Los Angeles, reductions in manufacturing in the early 1990s also reduced per capita use. Overall, the South Coast used nearly 450 000 acre-feet less water in 2005 than a decade earlier, despite having 2 million additional residents. As one study concluded,

> the temptation is to ... change the villain in California water policy from pool-loving residents of the South Coast to the urban and suburban residents of Sacramento, the San Joaquin Valley, and other inland areas. However, the urban sector as a whole accounts for just over twenty percent of water use in California, and utilities in virtually every region are working to reduce per capita use (Hanak et al., 2012b: 19).

Some 50–60 percent of residential water use in inland areas of the state is for *outdoor landscaping*, while indoor use rises with *single-family home* ownership – and as income grows, so do *outdoor and indoor* demands.

The regions where this tends to be true often do not employ tiered rate structures but apply uniform rates that charge the same amount per gallon (Hanak et al., 2012b; Hanak and Davis, 2009; Baumann et al., 1997). Moreover, for aesthetic or other reasons, some communities forbid measures that conserve water through, for example, removing lawns and replacing them with water-conserving landscaping.

WATER APPLIANCE EFFICIENCY

Mandated conservation measures that compel replacement of water appliances such as low-flow toilets and showerheads may also adversely affect low-income populations. While appliance retrofits can save water, especially in older urban areas, the cost of installation of appliances such as low-flow toilets and showerheads may burden poorer communities. Their installation in older urban neighborhoods may be arduous and expensive.

One strategy being employed by some utilities – in recognition of this access issue – is to partner with community groups to transform retrofitting programs into a form of community re-development strategy. In Los Angeles, the Department of Water and Power has partnered with environmental justice and other groups to install water-saving appliances, generate rebates, provide job training, and re-invest the funds raised for local schools, minority scholarships, drug treatment programs and other measures (Green, 2007). There is no consensus regarding the fairness criteria that should be applied to evaluating conservation innovations. In part, this is because many international agreements pertinent to water, as we have seen, view the provision of water as a basic human right. In effect, the moral default would appear to be, in most of these agreements, that water must be provided to people regardless of their ability to pay – but in accord with their needs. International experience would suggest that in societies where the state has effectively persuaded its citizens that providing clean, abundant and predictable supplies of water may require higher consumer rates, the acceptability of those rates can be made palatable if linked to actual performance criteria.

Indoor water conservation technologies have focused on repairing leaks and retrofitting or replacing household appliances to increase their water efficiency. Some examples would be installing water-efficient clothes washers, dishwashers, taps, water-saving showerheads or flow restrictors, as well as replacement of high-volume toilets (3.5 gallons or more per flush) with high efficiency or dual-flush models. Retrofits such as toilet dams, tap aerators, and low-flow showerheads can reduce indoor water use by 9–12 percent, while replacement of appliances can reduce water consumption 35–50 percent, based on studies from various cities in the U.S. and Australia (Inman and Jeffrey, 2006). Research from Seattle, San Francisco, and Tampa has shown that toilet and clothes washer replacement leads to the greatest savings for indoor appliances (Mayer et al., 2000; 2003; 2004). Toilet retrofits led to savings of 20 percent for Tampa and Seattle, and 6.6 percent in Seattle, due to a lower initial level of leakage. Leakage may lead to large water losses from a small number of homes. For example, it was found that 38 percent of homes were responsible for over 85 percent of the leakage in San Francisco homes (Mayer et al., 2003). Toilet replacement has shown large water savings in some cities, such as 37 percent in New York (Ostrega, 1996), and 44 percent in Salt Lake County, Utah (Mohadjer and Rice, 2004). The Sydney Water Corporation "Every Drop Counts" residential retrofit program, the largest residential demand management program conducted in Australia, found that participants had less demand compared to a control group and that the program achieved indoor water savings of 12 percent (Turner et al., 2004).

OUTDOOR TECHNOLOGIES AND PRACTICES

At the residential level, outdoor water use is typically not metered separately from indoor use. While some outdoor water use is attributed to cleaning (typically hardscapes and cars), it is primarily due to irrigation. Urban landscapes must be managed and irrigated sustainably in order to reduce demand on water supplies. Outdoor landscape species rely on irrigation, particularly in arid/semi-arid cities in the southwestern U.S., such as Los Angeles (Bijoor et al., 2012). It is likely that urban landscape irrigation accounts for a major portion of urban water use in arid/semi-arid regions (Bijoor et al., 2014; Gleick, 2003; Mayer et al., 2003). The conversion of cities into urban "oases" of lawns and trees increases the need for water exports. As the U.S. faced severe droughts in 2012 and 2014, urban irrigation has increasingly come under scrutiny, as outdoor uses are considered more discretionary than indoor uses. Urban irrigation not only strains water supplies, but also reduces surface and groundwater quality (SCCWRP, 2010). Increases in soil moisture as a result of irrigation increase greenhouse gas emissions from urban landscapes (Bijoor et al., 2008; Townsend-Small and Czimczik, 2010). However, interest in maintaining urban landscapes remains for several reasons, including atmospheric cooling, carbon sequestration, and air pollution removal (Akbari, 2002; McPherson, 1990; Mueller and Day, 2005; Nowak and Dwyer, 2007; Nowak et al., 2006), as well as aesthetics.

A number of technologies have emerged for making outdoor irrigation more efficient. Drip irrigation results in low volumes of water applied directly to plant roots, typically through a network of tubes. Rain sensors, also called rain shut-off devices or rain switches, pause the cycle of an automatic irrigation system during rain events. These are usually an inexpensive method of reducing excess irrigation and are often compatible with any irrigation system. Some cities, such as Cary, North Carolina, require that all new sprinkler systems be installed with rain sensors. Lack of uniformity in irrigation application may result in over-application, as landscapes are often watered so that parts receiving the least amount of water would still have their requirements met. Smart controllers, also known as "ET" or "weather-based" controllers, offer an alternative to automatic timer-based irrigation controllers commonly utilized by homeowners.

Automatic controllers tend to be inefficient because they apply irrigation irrespective of actual plant water need, which changes dependent on environmental conditions. Single-family homes in Los Angeles with automatic controllers have been shown to use 11.2 percent more water than those without (Chestnutt and McSpadden, 1991). Smart controllers are programmed to calculate plant water demand adjusted for landscape type based on weather and/or soil parameters, and apply water accordingly. A number of studies have shown that smart controllers may promote water and cost savings (Bijoor et al., 2014; Cardenas-Lailhacar and Dukes, 2012; Dukes, 2012; McCready and Dukes, 2011; USBR, 2008). Controlled experiments have shown water savings of 40–70 percent when using these devices, but some large real-world studies have shown savings of less than 10 percent (Dukes, 2012).

Smart sensors need proper programming and maintenance by the landscape manager, including accurate entry of horticultural information (for example, species type, soil type) and irrigation system parameters (for example, sprinkler output). These parameters usually need seasonal or more frequent adjustment. Although it is often assumed that smart sensors allow landscape managers to "set and forget," they do not eliminate the need for human intervention in landscape management. Smart controllers typically function best for irrigation systems that have a high degree of uniformity in water application. For this reason, smart controllers may be more effective when installed in conjunction with sprinkler nozzles that improve uniformity, such as rotary sprinkler nozzles. These nozzles operate by rotating a stream of water over the landscape instead of producing a mist in a single direction, thus improving sprinkler uniformity.

WATER CONSERVATION BEHAVIOR AND POLICIES

Supply-side management focuses on augmenting supplies through recycling water in increasingly smaller, decentralized projects, while demand-side management aims to affect water-consuming behavior. In some cases, water-use behavior may be influenced by education, outreach, demonstration and communication efforts made by water efficiency/conservation practitioners; in other cases, researchers have shown that water-conserving behaviors appeal to an individual's or group of individuals' values, beliefs, and understanding of the social norm (which is influenced by cultural, society, economic and environmental factors). Finally, we explore water conservation policies and management approaches, which have increased in scope and number as major climatic events have emerged in both Australia and California.

The majority of domestic water consumption in Melbourne and southern California is derived from the residential sector (Grant et al., 2013; Gleick et al., 2003; Hanak et al., 2012b). Thus, several studies have focused on this area to understand the dynamics of water-consuming behavior, including the overt or underlying elements driving water-conserving behavior. In an empirical study of more than 3000 Australians, Dolnicar et al. (2012) revealed that among several factors driving water-conserving behavior, two emerged as key: (1) a "high level of pro-environmental behavior," and (2) "pro-actively seeking out information about water" (Dolnicar et al., 2012: 44). Interestingly, a 2011 state-wide survey of single-family homes in California found that "only 16% of respondents agreed with the statement 'I conserve water mainly for environmental reasons', while 80 percent of respondents disagreed with this statement" (Aquacraft, 2011: 277). The same survey also found that respondents "had very little knowledge about how much water they use or how much money they spend on water"; researchers concluded that these results may "simply point out that there are more reasons for conserving water than just the environment (or economic) benefits" (Aquacraft, 2011: 277). However, it should also be noted that, on average, Australians use less water than Californians, even though they have a similar culture, economy, and climate (Cahill and Lund, 2011).

Another emerging approach that is being developed extensively in Melbourne and now starting to take hold in southern California is one that harnesses the power of community, by identifying community leaders to liaise between water institutions and the community and to help develop information-sharing networks among the local community.

Social Psychology and Water Use

Borrowing from social normative marketing theories, new outreach media and communication approaches have emerged to provide feedback to consumers on their usage relative to their neighbors. Companies that harness water consumption data, such as *WaterSmart*, share with individuals how their use compares with others within their peer group, defined as those households with similar household occupancy and landscape square footage. Additionally, the advent of social media, data-driven technology innovations, and the proliferation of personal mobile devices have provided individuals with direct and immediate access to data and have allowed them to manage their water consumption more effectively. Many of these tech-based opportunities have emerged in California due to the coincidental development of technology start-ups and the proliferation of smart meter technology. Technology companies such as *Dropcountr* have teamed up with local water agencies to provide individuals with updates about new water-related rebates, high-consumption and leak alerts, and helpful water conservation tips (Wang, 2014).

Interestingly, the social normative marketing approach yields positive results in both Australian and American contexts. The long-standing "Don't be a Wally with Water" marketing campaign began in 1984 to change Melburnians' attitudes toward wasting water by branding those who used more than their fair share a "Water Wally" (Melbourne Water, 2014b). This campaign again proved successful during the Millennium Drought (1997–2012) and was recently resurrected in the summer of 2013 by a major newspaper publication in Melbourne to temper a spike in water consumption during a heat wave (Dowling, 2013). Even Americans, a nationality typified by individuality and liberal personal property rights, are compelled by "a strong desire . . . to conform to accepted practices" (Hoffman, 2010).

Additionally, psychologist and water-use behavior change expert Laurent Lucas states that human competitiveness plays an important role in driving down water consumption, particularly if an individual knows that their neighbor is attempting to use less water than he or she. However, the social normative marketing approach is not without its critics, who argue that "neighbor comparison programs" are a "Big Brother" approach and inconsiderate of individual household needs (Hoffman, 2010). Outreach in the form of educational and information campaigns, as well as physical and behavioral water efficiency demonstrations, are major components of many water agency water conservation strategies, as well as the focus of recent academic studies (Dolnicar et al., 2010). Grant et al. (2013) noted that public education was arguably the most effective method utilized in Melbourne during the Millennium Drought to curb water consumption.

An experimental study conducted by Dolnicar et al. (2010) examined the public's acceptance of use options for recycled and desalinated waters. In the study, two groups were asked about their willingness to use each water source for a range of purposes; one group received information about the production process of each resource, while the other group received no information. The group receiving information about the processed water was more likely to accept a wider range of uses for it. The lessons gleaned from this study have important implications for policymakers and water conservation professionals seeking to convince the public to use a variety of water qualities for different purposes. Information/knowledge can provide a powerful motivator for adopting new household practices utilizing non-traditional water sources such as recycled water, stormwater, or graywater.

The findings of the Dolnicar et al. (2012) study and similar studies point to personal values and beliefs coupled with internal motivation as the factors that spur an individual to action. However, another powerful motivator influencing water-consuming behavior is the individual's perception of how their peers or neighbors consume water and whether the individual feels they fall within the social norm of water consumption.

OTHER EQUITY ISSUES

Recycled wastewater use presents several environmental justice and equity challenges that directly relate to low-impact development. Reused wastewater reduces the need for imported and diverted freshwater. In so doing, it alleviates pressures on supplies belonging to others – a major source of water disputes – and it reduces wastewater-generated pollution by alleviating the need to dispose of "dirty" water unfit for potable use in rivers and streams.

In considering whether the use of recycled water is fair, however, it is important to recognize three impacts. The first is the so-called "toilet to tap" issue: the perception that people are being asked to use water that quite readily has been taken from the wastewater stream and is immediately being used as tap water.

Second, greater reliance on water reuse and wastewater recycling to enhance and recharge groundwater resources and for prescribed potable and non-potable uses may actually encourage additional economic growth and water demands – while poorer groups are asked to consume recycled supplies. Opinion surveys find that it is often viewed by people as an indirect *subsidy* for unwanted additional population and residential and business growth (Boberg, 2005; Groves et al., 2008). And third, while a number of studies stretching as far back as at least the 1970s indicate that potable reuse is safe, doubts have arisen over safeguards. Important to the environmental justice debate is that perceptions of safety are associated with a sense of inclusion in decision-making. In less affluent areas, for those communities suffering from ongoing water issues related to environmental legacy (for example, abandoned hazardous waste sites, contaminated aquifers), when proposals for reuse are introduced, they are likely to arouse suspicion and widespread mistrust.

In short, conservation measures are characterized by *contending notions of equity*. On the one hand, as we have noted, those who can afford to should perhaps be entitled to greater use of water. After all, some would argue, they are paying for it and have earned the right to a certain kind of lifestyle. Conversely, those who sometimes pay more because of greater water use really are economically burdened. Their water use may not be calibrated at all by their income or lifestyle, but by greater needs prompted by factors over which they have little control, such as having larger families, special health needs, or the need to care for small children or the elderly.

In Chapter 3, we discussed how the Millennium Drought enabled Melbourne, and to a degree other Australian cities, the opportunity to enact and adopt several alternative approaches to water management that, in the long run, could signify a transition to a water-sensitive city. We now consider Melbourne's goals in the broader conceptual context of urban water management alternatives and their significance for other cities. Among other goals, Melbourne's aspirations are to become world leader in capture and reuse of stormwater runoff – for example, 2010 municipal demand (356 GL) < average annual stormwater runoff (463 GL). To accomplish this, the city's water planners seek to rely on several options to achieve this goal, including low-energy stormwater reuse schemes to (1) capture water before it becomes contaminated; (2) rely on low-energy processes for removing contaminants; and (3) treat water only to the extent necessary for intended use.

Melbourne's Experience

By the end of the Millennium Drought, Melbourne had undertaken or completed large centralized infrastructure projects as well as decentralized, locally based demand-attenuation and supply-augmentation projects. It had also conducted a number of information and public education campaigns. These projects and campaigns had various impacts on municipal water supply, cost, and reliability, but also on ecosystem health and function. To varying degrees, these measures were embraced by the public, playing an important role in local, state and national politics. Investments in the Wonthaggi Desalination Plant and in the North–South pipeline were motivated by a desire to provide Melbourne with a more reliable (drought-proof) supply of potable water.

While these projects did not contribute to Melbourne's water supply during the Millennium Drought, projections suggest they may be needed in the future. Melbourne's experience with wastewater recycling is more nuanced. The use of recycled wastewater for local agriculture increased dramatically during the Millennium Drought, but for a variety of reasons the initial uptake did not last. By contrast, urban and residential use of recycled wastewater appears to be on a long-term increasing trend. Rainwater harvesting was embraced by the public, with positive impacts on both Melbourne's potable supply and possibly ecosystem health. The Millennium Drought also spurred interest in stormwater harvesting, in part because of its potential to reduce a large fraction of the city's long-term water needs. Government programs to restrict water use, improve water efficiency, and educate the public were both highly effective and relatively low cost. Indeed, the success of Melbourne's water conservation programs kept the city from running out of potable water during the Millennium Drought.

Among supply augmentation approaches, recycled water use in agricultural irrigation achieved the greatest gains by volume, largely from use in the Werribee Irrigation District, as a result of historic low river flows and the ban on groundwater use. Since the intended use of recycled water was only to supplement river water, and the use of recycled water was driven by the lack of alternatives, recycled water use dropped after the drought broke. Moreover, sustaining high levels of water reuse would have been politically and economically difficult, partly because of increasing soil salinity that would have decreased agricultural productivity over the long run. Because of its high cost, there are currently no plans to build a salt reduction plant (Low et al., 2015).

Lessons for the Water-Sensitive City

Melbourne's experience with the Millennium Drought offers many lessons, both positive and negative, for other similar cities. For one, the severity of the Millennium Drought afforded Melbourne a window of opportunity to adopt supply-side and demand-side measures that in normal times may have proven very difficult, if not impossible, to adopt. One lesson for other cities is that major droughts, if serious enough and long-lasting enough, create opportunities for policymakers, as well as posing challenges. This is only true, however, if policymakers choose to act on these opportunities by exploring diverse and multifaceted programs. Investments in the

Wonthaggi Desalination Plant, the North–South pipeline, and the alternative water sources entailed significant risk for the politicians who made these decisions, particularly if the drought had proved to be short lived.

The Millennium Drought also encouraged public discourse on demand-side measures and low-impact development, activities whose long-term benefits are significant. For example, there was a high rate of compliance with, and widespread acceptance of, water restrictions, despite their severity and long duration. There was also a high uptake of rainwater tanks in the residential sector, showing the public wanted to do "their part." In addition, the Target 155 campaign was highly successful, reflecting the population's willingness to voluntarily alter their water use behavior for the greater good. This public support was due, in part, to an intense media campaign aimed at communicating the dire nature of the water supply situation (water volumes in the reservoirs were updated on billboards and on the front pages of the local newspapers) and suggesting ways of reducing demand. Melbourne's experience with the Millennium Drought unequivocally demonstrates that water conservation and other "soft-path" approaches can be an effective approach to fighting drought and improving long-term urban water sustainability. Furthermore, Melbourne was able to save water without resorting to the use of water pricing as a drought-response measure. The latter approach could have alienated sections of the community because of inequity (perceived or otherwise) between those who can afford to be wasteful and those who would suffer undue financial hardship.

A challenge that many cities, including Melbourne, face is the ability to sustain conservation efforts following the end of a drought event, especially given the initial cost of conventional conservation measures. A wide range of water conservation initiatives were implemented that varied greatly in cost effectiveness and amount of water saved, but a smaller number of well-chosen initiatives might have been more effective, less costly, and easier to implement. While Melbourne's investment in alternative water supply means that the city will be more resilient in the face of future droughts, pressures brought on by government administrative reorganization, public "fatigue" with continued sacrifice, and the practicality of financing these initiatives constitute a continuing challenge (Low et al., 2015).

In sum, evaluating urban water innovations compels decision-makers to pose – and seek to answer for themselves – a number of important questions. These include: how easily can innovations be integrated into urban landscape at multiple scales? Can the public and decision-makers embrace benefits of approaches that avoid channel modification, remove contaminants, and avert other adverse impacts? Can regulatory frameworks be developed to encourage water substitution and minimize human health risks – while encouraging private investment?

And finally, can advantages of centralized/decentralized governance be combined? We know that decentralized solutions increase resilience and adaptability, and that centralized governance (for example, in Australia) allows for region-wide performance targets that are difficult to achieve under weak centralized control. In short, governance issues are a key to introduction, management, and operation of effective urban water innovations, as we discuss in the next chapter.

9. New forms of management and governance for urban water sustainability

Predicting how cities will manage water problems in the future requires understanding how they have managed water in the past. Our earlier discussions in Chapters 2 and 3 considered, in part, this very issue. We need a deeper, more sophisticated explanation for urban approaches to water management that helps to explain why certain approaches to infrastructure development have been pursued, and that can help us understand the prospects for a water-sensitive city in the future.

This chapter begins with a discussion of *three* theories that help explain how cities foster water conflicts with their neighbors and among themselves. Most importantly, these theories help explain how urban water is controlled, under what conditions equity is jeopardized and why cities aspire to grow, and why additional infrastructure to draw water from remote regions is pursued. They also help us understand what changes in governance may be required to introduce innovations that can bring about a water-sensitive city. We then move on to discussions of novel governance arrangements that can foster and sustain this objective: especially, the kinds of governance systems that can promote green infrastructure, conservation innovations, and supply-side novelties.

URBAN GOVERNANCE AND WATER I: THE CITY AS GROWTH MACHINE

The *city as growth machine* approach has been embraced by a number of urban planning theorists and social scientists. Also called the "treadmill of production" model, it advances the notion that cities are engines of capitalist production that seek to amass capital on behalf of powerful elites. In the process, they transform resources into commodities that are bought and sold, and they generate socially exploitive, environmentally unsustainable conditions for their surrounding environment.

Initially popularized by Marxist theorists, this approach has been embraced by a number of urban planners and social scientists (Schnaiberg and Gould, 1994; Molotch, 1976; Logan and Molotch, 1987).[1] More recently, this approach has come to be embraced by so-called "deep ecologists" who view cities and urban life, generally, as products of capital-forming elites who seek to transform the environment into a controlled artifact for individual material aggrandizement (for example, Sessions, 1995). For urban water policy, the significance of this theory is that it predicates that cities invariably pursue patterns of water resource development that view infrastructure development and importation as means to create and market goods produced in cities. In the process, resource depletion and pollution inevitably result. Today this same pattern – found extensively in developed, industrialized societies – is being replicated in developing nations, particularly in "megacities" found in India, China, and parts of Latin America and sub-Saharan Africa, with, as we have glimpsed, an emphasis on large-scale conventional infrastructure. Moreover, as in cities in Europe and North America in the nineteenth century, public participation in decision-making is traditionally confined to knowledgeable technocratic elites. Workers feed the "growth machine" but rarely benefit from it, except as co-opted victims of its exploitative ways; in addition, all forms of land use development deplete the environment irreversibly – without radical political change. Little thought is given to preservation except insofar as depletion threatens further growth.

The *growth machine paradigm* may help us understand both the continued reliance on large-scale infrastructure as well as some of the challenges that face green infrastructure and other components of the water-sensitive city approach. The theory offers a compelling means of explaining why cities resist limiting growth, and in turn water use, as well as why cities historically have tended to pursue "exclusivist" means of water decision-making, ignoring, or even actively opposing the views of external groups whose interests are threatened by diversion projects, for instance, as we discussed in Chapter 2 in our discussion of Los Angeles and New York. These cities nearly exhausted available supplies and then sought additional water elsewhere by employing a combination of private investment in new infrastructure and political alliances with outside agencies.

The city as growth machine theory also helps explain why the path to seeking solutions to urban water problems in many large cities is heavily trod by elite decision-making. As we have seen, entrepreneurial leaders, high-ranking elected officials, and other powerful urban elites have tended to dominate in decisions regarding water resources development in cities. This condition is also becoming characteristic of water policy in developing nations (Olivera and Viana, 2003; Crow and Sultana, 2002).

This theory has one potentially glaring deficiency, however: it oversimplifies the dynamic nature of many urban water problems. The driver of cities' thirst is not always capitalism, out-of-control market forces, or commodification of water. Policy failures caused by urbanization are often driven by misguided government decisions to divert water and buy and sell water rights in order to "move" water from uses deemed low in value to those seen as more advantageous for development. *Complicity* between capitalist elites and government has often been at the root of these bad decisions, and the deeper cause of elite dominance is a policy-making process that impedes democratic participation and inhibits bureaucratic accountability. These conditions can as readily be found in state socialist systems as in capitalist ones, and make resistance to water innovations ubiquitous across various types of societies (for example, Barlow and Clarke, 2002).

URBAN GOVERNANCE AND WATER II: ZERO-SUM CONFLICT AND WATER

The *zero-sum conflict* theory predicates that understanding urban water management requires an examination of the ways in which race, ethnicity, gender, and other demographic factors affect access to water rights, bias allocation decisions toward those who are well-off, and lead to unfair and insecure access to water for others. In effect, any resource development policy that produces gains for one key group or interest produces a loss on the part of another – hence the expression "zero-sum."

Some political scientists and other advocates of this theory contend that by virtue of their large populations and ethnic and lifestyle diversity, large cities engender especially fierce competition between groups, particularly during periods of growth. This view is embedded in environmental justice critiques of water and other resource management issues (for example, Minkler et al., 2006; Pellow, 2002; Hurley, 1995). Competition among different classes, races, and ethnic and nationality groups produces further competition over resources, and efforts to seek special protection from the deleterious effects of environmental nuisances and hazards. Furthermore, exposure to differential hazards and environmental nuisances is viewed as a form of virulent injustice that may generate fierce resistance. This approach also claims that cities seek to manage water to support growth more than worrying about its equitable allocation among these diverse groups; priority tends to be accorded to supply- rather than demand-side options among urban decision-makers. Demand-side management embraces the virtue of doing more with less water, of saving money and energy, and of seeing cities *not* as *growth machines* that must be constantly fed to sustain jobs, but as centers for experimentation to reduce sprawl, conserve water *and* land, encourage "livability," and reduce our environmental footprint.

This theory has drawn the attention of many international forums on water rights (World Civil Society Forum, 2002; Coulomb, 2001), but has also come under criticism for allegedly failing to recognize that not all water use conflicts produce such sharp contrasts between benefits allocated to different groups. Water markets, for example, have the potential if properly managed – to facilitate reallocations in ways that can benefit both those who sell or lease water, as well as those who buy or rent it.

However, successful urban market exchanges require protection of environmental and social values and aspirations as well as efficiency, and this has not always been the case in urban water markets that move water from rural, agricultural centers to cities (Ingram and Oggins, 1992). In short, a challenge facing the zero-sum approach is being able to prescribe the kinds of structural changes needed to ensure that urban water decision-making becomes more transparent, open, and accessible to disadvantaged groups – and the tools and approaches that need to be introduced to make this possible.

URBAN GOVERNANCE AND WATER III: METROPOLITAN NATURE AND ITS IMPLICATIONS

Early in this volume – beginning in Chapter 1 – we suggested that one of the most significant challenges facing prospects for a water-sensitive city is the implicit failure of urban planning efforts to introduce rationally based programs to forestall sprawl, low-density suburbanization, and the trend toward "edge" development and its accompanying infrastructure costs. Large cities, more often than not, seem to confirm the inevitability, if not desirability, of the city as "growth machine."

The *city as metropolitan nature* is an approach to understanding the modern city which has been increasingly embraced by historians, political scientists, and urban planners and predicates another view of how cities actually grow – as well as how they relate to and interact with their natural environments. Also termed the theory of "second nature," this approach contends that in their growth and development, cities end up as they do not because of the failure of planning but because natural *and* human systems are dynamic. A new, altered nature is brought about by the mix of "first patterns" of development (those following early settlement), together with "first nature" (the pristine environment upon which these settlements are founded). *Second nature* cannot be understood apart from the evolution of environment and society together. Moreover, it would be incorrect to conceive of the result as simply degraded nature. It is more accurately understood as a dynamically transformed environment which generates distinctive patterns of beliefs on settlers and their progeny, altered species of flora and fauna, new kinds of property ownership and market relations, and altered climates (Cronon, 1992).

For urban water policy, this approach reminds us that, at their best, cities are an artifact – a product of powerful interactions between nature and urban dwellers that results in a regenerated hydrological system. Phenomena such as urban heat islands and "regional aridity and the struggles to control water" are apt metaphors for transformation from first to second nature, as is the creation of open space, community gardens, parks, and even sprawling housing tracts made to look like "oases" (complete with imported palms, flower gardens, fountains, and the like). The transformation of environmental beliefs, while less easily measured, are no less apparent (Hise and Deverell, 2005). Moreover, great cities seek from their inception to modify, harness, and transform waterways, both within their territorial boundaries and outside of them in ways that accrue to their growth and development. We saw this is Chapter 2 with Athens, Rome, and Barcelona, as well as with Los Angeles, New York, Mexico City, Tokyo, and Melbourne. In Los Angeles – where the ethic of water harnessing and control arose early on in its development – efforts were expended to accommodate the growing demands of an agro-urban economy for a steady water supply (a combination of municipally managed *"zanjas"* and pueblo legal claims on remote water sources in the San Fernando and San Gabriel valleys, including groundwater). This "water ethic" sought to restrain the basin's vacillating rivers from causing property damage through publicly sponsored engineered channelization (Hundley, 2001; Gumprecht, 2001; 2005). Similar attitudes, initially, were sought in other water-challenged cities as well.

In short, this approach helps us understand the complex social ecology of water use, and the environmental as well as social changes brought about by efforts to develop, use, and exploit water in modern cities – as well as to restore riverine environments in cities to attract gentrified redevelopment. Imperial attitudes toward neighboring sources of water only change when a growing adaptive management paradigm is forced upon large cities by external environmental issues (for example, chronic drought, climate variability), economic pressures (for example, competing water uses), and political resistance (for example, efforts to protect or restore environmental amenities in, or in areas adjacent to, cities).

How can we move from one domain to another? Moreover, how can cities become more water sensitive, given the factors these three theories – especially the theory of second nature – help to illuminate? The answer lies in identifying conditions that permit a "paradigm shift." While we earlier saw examples of such shifts, historically, when cities transformed from earlier epochs of infrastructure development to another (Chapter 4), we emphasized the roles of culture, economic development, technology, and public input and demands.

While these factors will remain important to a "shift" toward a water-sensitive city, other changes also must occur. With respect to governance, several factors must be encompassed. These include overcoming bureaucratic inertia, satisfying private developer concerns with returns-on-investment, surmounting the tendency toward fragmented and piecemeal water resource planning, and including civil society groups in major decisions (Van de Meene et al., 2011). Polycentric governance is one management approach that has been proposed as a means for meeting these challenges.

GOVERNANCE FOR WATER INNOVATIONS: POLYCENTRIC GOVERNANCE AS SOLUTION

The concept of polycentric governance originated in observations of how locally incorporated communities, typically found in U.S. metropolitan areas, improvise solutions to cross-jurisdictional problems – particularly as regards water – through contractual agreements. Subsequent studies examined the use of polycentric governance in managing "common pool" resources such as freshwater, fisheries, and forests, and implementing public safety programs in entire regions (Ostrom, 2010).

The appeal of polycentric governance among some water innovation enthusiasts lies in the opportunities it affords for building confidence among stakeholders regarding ways to overcome uncertainties and possible risks associated with various innovations. These opportunities begin with collaborative mechanisms to manage risks in ways that engender public trust and confidence.

Polycentric governance features multiple, relatively independent centers of power; there is greater opportunity for locally appropriate institutions to tightly monitor developments within a policy area – including emerging risks or hazards – and to introduce locally accessible trustworthy approaches to their mitigation. Moreover, its advocates claim, such local monitoring affords effective early warning, "safe-to-fail" interventions (in other words, policies that – should they not prove immediately to reduce risks can be quickly replaced by newer, more adaptive innovations). In addition, by encouraging open communication and deliberation to build trust and shared understanding among diverse stakeholders, polycentric governance creates opportunities for social learning in places and scales that better match the spatial context of problems (Lebel et al., 2006).

Polycentric governance for water innovations would ideally entail a collaborative effort in which: (1) knowledge and innovation developers are able to systematically straddle disciplines (cross-disciplinary collaboration is intrinsic to this enterprise as we have discussed – less so for innovation within individual scientific domains); (2) these innovators are prepared to utilize social networking media and other communication innovations in order to foster "flat," cooperative and so-called "leaderless" movements of partners and associates; and (3) would seek to bring together all the actors, rules, conventions, processes, and mechanisms concerned with relevant information on how to conjointly manage supply and demand-side innovations (Renn and Roco, 2006). These can be achieved by providing a platform for public–private partnerships to build public trust and confidence (Smith and Stirling, 2010). Let's look at some possible examples of where this model seems to be emerging.

Melbourne and Polycentric Innovation for Drought Response

During the Millennium Drought, Melbourne and the state of Victoria achieved the introduction of numerous demand- and supply-side innovations, as we have seen. Beyond these technical approaches and behavioral changes, however, was the ability to transform the ethos of decision-making for water in a large metropolitan region from a traditionally fragmented, multi-jurisdictional approach to one predicated on a culture of community engagement and societal innovation which permitted adopting solutions that could be generated from suites of multiple options – including low-impact (or green infrastructure) alternatives (see Figure 9.1).

In short, administrative and governance changes both encouraged, and helped sustain, these innovations. In fairness, while many of these administrative innovations lagged *behind* technical ones in their introduction, nevertheless, the city's capacity for introducing technical innovations was directly facilitated by institutional and *governance* factors that permitted an integrated policy response to drought. Melbourne's water companies are constituted under the Victorian Water Act of 1989. State Ministers of Water, Environment, and Health, and Treasury collectively oversee the water sector.

A Department of Environment and Primary Industries (DEPI) supports the Minister for Water and the Minister for the Environment, while an Essential Services Commission regulates the water sector. Retailers must submit plans to the Commission to justify rate increases, while dividends

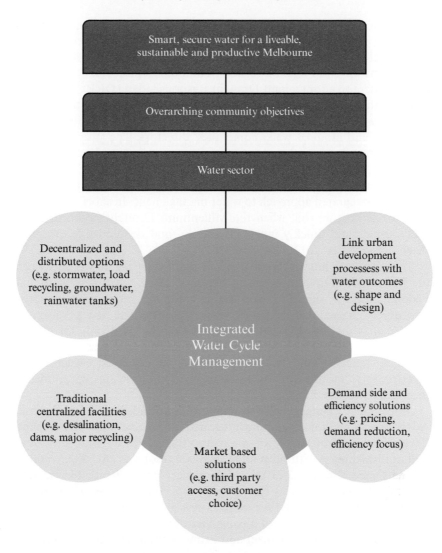

Source: Ministerial Advisory Council for the Living Melbourne, Living Victoria Plan for Water (2011).

Figure 9.1 Water innovations in Melbourne: interconnections and role of governance

are paid annually to the State Treasury. Unlike the "fragmentation" often typical of most water utility management patterns, however, a clause in the Water Act allows the Minister for Water to issue a Statement of Obligations (SoO) in relation to water companies' performance, and also requires them to adopt a joint Drought Response Plan (DRP).

While the DRP specifies various levels of water restriction based on water storage levels, the Minister for Water makes or lifts restrictions with the advice and input of water companies' directors. The latter, as well as DEPI, also influence other drought responses, reflecting Victoria's deep interest in projecting strong leadership, while the Council of Australian Governments' National Water Reform Framework of 1994 promotes a nationally integrated approach to water management. In short, this framework helped assure that when the Millennium Drought began in 1997, Melbourne could quickly introduce supply- and demand-side measures (Melbourne Water, 2013; Low et al., 2015).

Amplifying the aforementioned problem of policy lag, it was only when the Millennium Drought began to break, in 2010, when new State and Federal elections were held, and the new administration which entered into office adopted a more novel policy platform. The new government decided that the North–South pipeline was to be used only for critical human needs (defined as storage volumes in Melbourne's reservoirs falling below 30 percent); and the desalination plant discussed earlier would be effectively "mothballed" as an example of the previous government's mismanagement and waste.

Other policy changes included appointing an independent Ministerial Advisory Council for developing integrated alternative water sources, and the establishment of the Office of Living Victoria (OLV) (May 2012) – as well as furnishing it with substantial financial resources (AU $82.5 million from its inception to 2013/14) (Melbourne Water, 2014b; Siriwardene et al., 2011). These reforms encouraged harvesting and use of rainwater, stormwater, and wastewater recycling as new water sources for Melbourne. Since July 2014, the Office of Living Victoria has been reorganized into the Department of Environment and Primary Industries and is no longer an independent organization. Because these investments and decisions are largely political, it is unclear if the OLV will continue to enjoy this level of support, and if it will continue to prioritize, for example, the development of stormwater harvesting. More generally, there is an overarching challenge associated with estimating the costs and benefits of different water supply and demand reduction approaches that properly account for economic factors, environmental health, public acceptance, and livability. To the best of our knowledge, there is no comprehensive study on the price (including externalized costs) of distributed and centralized water management approaches compared to the water volumes gained or saved (Low et al., 2015).

From the standpoint of polycentric governance and water innovation, it may be stated that the Millennium Drought afforded Melbourne a window of opportunity for supply- and demand-side measures that in normal times may have proven very difficult, if not impossible, to adopt. One lesson for other cities is that major droughts, if serious enough and long-lasting enough, create opportunities for policymakers to take dramatic governance as well as technical actions. The ability to take advantage of this window of opportunity depends, however, on the willingness of decision-makers to engage the public, as well as institutional conditions which encourage adopting innovations. Perhaps ironically, such a window of opportunity also requires the security provided by investment in some "hard path" alternatives which reinforce the public's confidence that diverse and multi-faceted programs and options are being pursued by civic officials just in case they are needed.

California: Is Polycentrism Possible?

California exemplifies a long-standing pattern of water policy-making that – while ubiquitous in the U.S. – is especially dominant in the West. An "old way of doing business," as one former official has phrased it, for water governance in California is by way of a diffuse and disparate govern-ance structure comprised of numerous institutions having a single-purpose focus. This has been the prevailing pattern of water governance since the mid-1800s and has dominated management of the state's water resources – both urban and rural.

In fairness, this fragmented approach made sense, at least initially. Fragmented governance in the water sector reflects the fact that societal demands with respect to water first arose in the context of resource extrac-tion (principally, gold and silver mining). Later, governance was adapted to other purposes for which pre-existing policies, institutions, and laws were less than adequate – including reclaiming of wetlands and meeting the growing needs of agriculture and cities. Taken together, these needs promoted, rather than mitigated, the growth of water management and water regulation silos (California Department of Water Resources, 2013a).

While slow in coming, efforts to consolidate these silos, and to surmount fragmentation, have occurred – and in substantive areas where polycentric governance could be most effective. With respect to urban water needs, for instance, the Integrated Regional Water Management Act of 2002 (SB 1672) allocated $1 billion from the Water Quality, Supply and Safe Drinking Water Projects Bond of 2002 (Proposition 50) to spur integrated water management projects throughout the state. The Act enshrined in law the ability to capture and use rainwater harvested from rooftops for non-potable and prescribed uses.

Further augmenting this effort, and departing from Western states' long-standing tradition of making it illegal to capture and use precipitation based on the prior appropriation doctrine, the California legislature enacted the Rainwater Capture Act of 2012 (2012 Cal. Stats. ch. 537, Sec. 2), which exempts the capture and use of rainwater from rooftops from the State Water Resources Control Board's (SWRCB) permitting authority over water appropriations. This Act affords residential users and private and public entities with a new source of onsite water supply, which could reduce reliance on potable water for landscaping needs and provide a recharge benefit to underlying groundwater aquifers. Other obstacles, including public acceptance of stormwater harvesting, still need to be overcome.

Prior to enactment of the Act, the SWRCB required all would-be appropriators to apply for and obtain a permit to appropriate water from any source, including water falling in the form of precipitation. Now, however, the use of rainwater – defined as "precipitation on any public or private parcel that has not entered an offsite storm drain system or channel, a flood channel, or any other stream channel, and has not been previously been put to beneficial use" – is not subject to the California Water Code's SWRCB permit requirement (California Water Code §§ 1200 et seq.). Relief from the permit requirement enables residents, private businesses, and public agencies to create new onsite water supplies to meet landscaping needs, thus decreasing the use of potable water to meet those needs while acknowledging existing water rights, "and the desirability of not impairing the authority of water suppliers to protect the public health and safety of public water supplies as authorized by California law" (Davis and Slater, 2013). The law also recognizes the state-wide "20 × 2020 goal," which aims to achieve a 20 percent reduction in urban per capita potable water demand by 31 December 2020 (2012 Cal. Stats. ch. 537, Sec. 2). This goal may become more realistically achievable given pressures during the current drought to dramatically reduce water uses.

New policy goals (for example, to transition from cheap water provision to grow a metropolis to satisfying the "co-equal" goals of public safety, environmental stewardship, and economic stability) that "deliver multi-benefit programs and projects across watershed and jurisdictional boundaries" (California DWR, 2013a: 4) may, over time, trigger change in strategies to achieve sustainable water system management by promoting integrated water management.

One challenge with California's current governance system – as we have alluded to – is that most agencies have been established to focus on a few objectives (for example, supplying water, protecting/regulating water quality, or protecting/enhancing fish and wildlife habitat). Objectives outside of one agency's area of responsibility or window of authority are either assumed by a different agency, or not addressed at all. Some water resource managers cite limitations within their current authorities and organizational missions that prevent or hinder them from engaging with others to collectively address broader water management objectives. Also, in some cases, certain sources of funding can only be used to satisfy specific (and sometimes narrowly defined) objectives.

Concerns regarding agency scope-of-authority and restrictions on funding as inhibitors of collaboration warrant further exploration. While the case can be made, and is beginning to gain traction, that more innovative and flexible governance structures, institutions and funding mechanisms are needed to support sustainable urban water management, it will take time to implement durable reforms. In the meantime, state, federal, tribal and local water officials need to work together to embrace incremental, yet effective, improvements toward greater polycentric water governance "that is permissible within the current structure and authority of the various governing agencies" (California DWR, 2013a: 7). Financial incentives are a good way to spur greater integration.

In 2013 the California Department of Water Resources recognized 48 integrated water resource management (IWRM) planning areas, whose participants were working together to manage their water resources more efficiently. Plans spanned 87 percent of California's geographic area and 99 percent of its population. IWRM has evolved into a major component of the California Water Plan (also known as Bulletin 160), which is updated by DWR every five years to serve as a roadmap of the state's water future. An excerpt from the 2005 Water Plan, the first upgrade published following passage of the Act, stated that "Integrated regional water management is the future for California because it will help regions diversify their water portfolio strategies and get the most from local, state and federal resources and funding" (Water Education Foundation, 2013).

Ironically, unresolved conflict between Los Angeles and Inyo and Mono Counties, and the EPA, to restore some 62 miles of the lower Owens River, to "re-water" portions of Owens Lake, and to allow the return of flows through Owens Gorge, and to restock bluegill, largemouth bass, fingerling trout, and other aquatic species – an effort that initially began in 1994 (see Chapter 2) – may have hastened greater IWRM. Los Angeles will ultimately receive 18 503 fewer cubic meters (15 000 fewer acre-feet) of Owens Valley water each year, reducing its reliance on the Owens

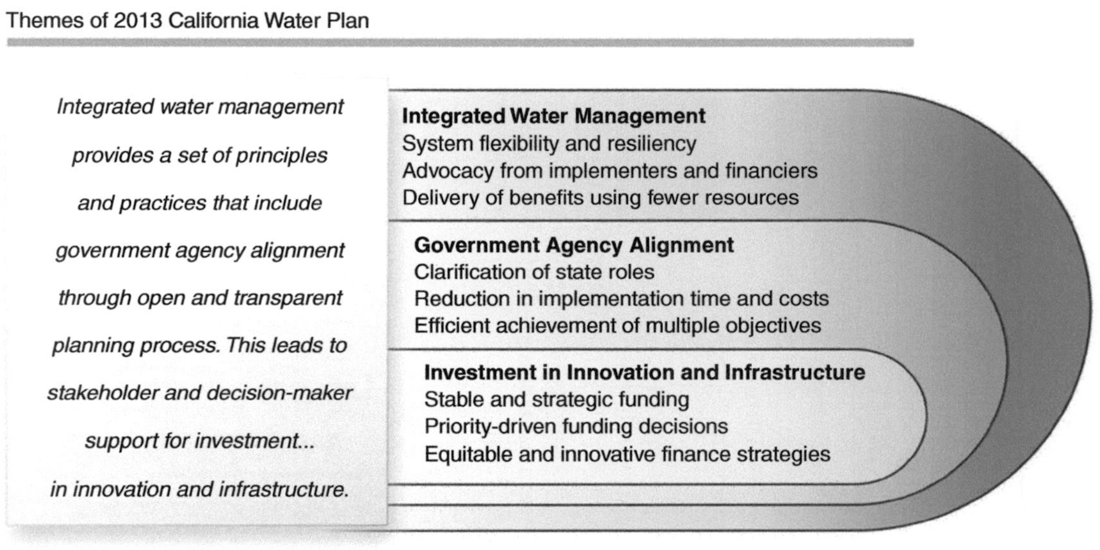

Themes of 2013 California Water Plan

Integrated Water Management
System flexibility and resiliency
Advocacy from implementers and financiers
Delivery of benefits using fewer resources

Government Agency Alignment
Clarification of state roles
Reduction in implementation time and costs
Efficient achievement of multiple objectives

Investment in Innovation and Infrastructure
Stable and strategic funding
Priority-driven funding decisions
Equitable and innovative finance strategies

Integrated water management

provides a set of principles

and practices that include

government agency alignment

through open and transparent

planning process. This leads to

stakeholder and decision-maker

support for investment...

in innovation and infrastructure.

Source: California Department of Water Resources (2013a).

Figure 9.2 Governance and water innovations for California urban water

Valley from some 35 percent of its total imported supply to approximately 18–20 percent (Linder, 2006; Hundley, 2001).

Of at least equal, if not greater, significance is the change in decision-making process by which these policies are being implemented. A Collaborative Aqueduct Modernization and Management Plan, or CAMMP, led by LADWP, the California Department of Fish and Game, and two environmental groups – California Trout and the Mono Lake Committee – has been undertaken to determine the means by which aqueduct operations can best be modified to facilitate changes in streamflow that can satisfy environmental restoration needs on the one hand, while continuing to provide water to Los Angeles. Thus far, extensive data gathering, analysis, and drafting of prescriptions have been conducted, and the effort has entailed far more cooperation among protagonists than in the past (McQuilkin, 2011).

Finally, and in parallel, while these environmental restoration activities are occurring, another collaborative effort has been conducted, off and on, regarding Native American water rights in the Owens Valley. Several Paiute Indian tribes lost their land and water rights in the region following white settlement in the mid-nineteenth century, and well before the aqueduct was built. A partial restoration of water rights occurred in 1908 following a pivotal Supreme Court case – *Winters* v. *U.S.* – which "explicitly affirmed water rights on Indian Reservations" by, in effect, setting aside correlative water rights on these reserved lands (Burton, 1991). All in all, efforts to overcome fragmentation, to collaborate across sectors and regions and among different groups – and to formulate over time a new environment for urban-related water governance are taking place in California. The final form these efforts take is still very much up for grabs.

New York: Polycentrism and Source Water Protection

As we discussed in Chapter 2, New York City, like many of the world's great metropolises, sought to acquire, control, and place under unified governance its sources of public water to assure dependable quality and available supplies. As we saw, efforts straddling the better part of a century were undertaken to secure much of the supply of the Catskill and Croton watersheds and, eventually, much of the flow of the Delaware system. Fortunately, the storage reservoirs built by the city – surrounded by forests that naturally filtered contaminants and forestalled erosion – helped the city to ensure its residents a pristine supply at relatively low cost.

By the 1970s, however, these assets were no longer able to ensure both objectives satisfactorily. Water quality in the Croton and Catskill watersheds began to deteriorate as a result of contamination from sewage outfalls, leaky residential septic systems, agricultural runoff, and land cleared for residential development. The most significant issues that arose were: (1) sediment problems or turbidity within the Catskill Watershed, which can transport pathogens and interfere with the effectiveness of water filtration and disinfection; and (2) excess nutrients, particularly phosphorus. The former can generate algae blooms that cause serious odor, taste and color issues, while excess phosphorus can cause eutrophic water conditions and increase carbon. Moreover, this water, mixed with chlorine, can result in the formation of "disinfection byproducts" suspected of being carcinogenic (New York State Department of Environmental Conservation, 2010b).

After years of study, environmental protection officials in New York City – and state officials representing the Department of Environmental Conservation – concluded that there were two feasible options to forestall threats of federal intervention, by the Environmental Protection Agency, to institute more strenuous remedial measures. The first was to build an artificial filtration plant, the city's first, at an estimated cost of $8–10 billion, with an annual operating expense in the vicinity of some $360 million. The second option was to restore the Catskill/Croton watersheds through a combination of land purchases, compensation of existing private property owners for growth restrictions (for example, conservation easements), and subsidies for septic system and other improvements. The city chose this much less expensive option (at a total cost of approximately $200 million) – paid for through the sale of municipal bonds (New York State Department of Environmental Conservation, 2010b).

This second option – now known as the New York City Watershed Protection Plan – was as much governance innovation as technical achievement. It has effectively assured compliance with federal drinking water standards and is delaying the need for a filtration plant, and is based on explicit, legally binding, and multi-jurisdictional agreements. These consist, first, of a Filtration Avoidance Determination (FAD) agreement, and second, a Memorandum of Agreement (MOA) that was concluded in January 1997 between several federal, state, New York county and city agencies, as well as various educational and non-profit organizations and watershed coalitions. Both agreements provide regulatory oversight, perform environmental monitoring, protect water quality, educate the public, communicate about issues pertaining to pollution and watershed stewardship, and provide funding and other assistance to watershed communities (Westchester County Department of Planning, 2009: 2–26).

This partnership acknowledges the common interest of both public and private entities – in the city and within the two watersheds – in abating pollution through working together, especially given the limited power of any single entity to abate non-point pollution. Unlike the Los Angeles case, where collaboration on environmental quality issues initially emanated from an adversarial clash of interests, this partnership came about more amicably, while its composition has been similarly diverse. Members include New York City agencies, upstate communities in the twin watersheds, the EPA and other federal agencies, the New York State Department of Environmental Conservation (DEC), other state agencies, and various environmental groups.

One explanation for this comparatively amicable partnership is political realism: most watershed communities would have been adversely affected had New York City been forced to build a drinking water filtration system. This is so for two reasons: (1) the plant would have been paid for by all water users (and, in all likelihood, by regional taxpayers); and (2) the state – if not the city itself as eminent domain tenant – would have been forced to impose more onerous land-use controls over the watershed if a partnership had not been formed. In effect, the indirect threat of having to pay for a water filtration plant was exactly the incentive needed to collaborate. Moreover, the choice of a multi-party partnership best suited the goals of all protagonists. It offered a viable, effective solution at manageable cost and through largely voluntary action (Croton Watershed Clean Water Coalition, 2009). However, given continued growth in rural areas throughout the region, and continued problems with turbidity, it has been necessary to revisit this plan.

In 2004, the city began construction of a $2 billion underground filtration plant in Van Cortlandt Park, Bronx, designed to filter water from the Croton system, which is scheduled to be completed in 2012. It has also continued to acquire sensitive lands in the Catskills/ Delaware watersheds to further buffer their reservoirs from contamination, and thus, to continue in compliance with the state/EPA approved FAD agreement (New York City Department of Environmental Protection, 2010).

Local collaboration was abetted to some degree by federal and state government action. For the former, EPA intervention forced Los Angeles to rectify the condition of Owens Lake (and thus, indirectly, also to improve the condition of other valley watersheds affected by adverse flows). Ironically, violation of the Clean Air Act (not the Clean Water Act) forced the city to work with state agencies, local valley officials and intervenor groups. For New York City, it was the *threat* of EPA (and state regulatory) intervention under the Safe Drinking Water Act (the Croton and Catskills are, after all, potable water sources) which compelled the city and its neighbors to collaborate to avert further sewage and non-point runoff contamination of the region's reservoirs.

CONCLUSIONS – CHALLENGES TO POLYCENTRISM: PUBLIC ENGAGEMENT

A number of issues must be confronted in order to optimally fit polycentric governance to the management of urban water problems. These include: (1) investing in appropriate governance innovations that permit durable decision-making frameworks and information sharing – particularly regarding potential risks; and (2) facilitating meaningful public participation mechanisms, including facilitating dialogue and permitting direct engagement by the public in governance.

The polycentric governance literature speaks to the need for multi-level governance able to deal with "different scales of market and government failures." It also refers to the ability to draw upon the experiences of cross-jurisdictional governance units in other areas of polycentric governance (for example, economic development zones, tribal districts, school districts, water utility districts, charter cities, trade and monetary zones), by arguing that specialized units for provision, production, financing, coordination, monitoring, sanctioning, and dispute resolution are needed to address issues of public trust and confidence (Araral and Hartley, 2013). Other polycentric governance scholars note the importance of encouraging active participation of local users – especially in managing private, locally managed, or even state-governed common pool resources – through, among other means, matching governance considerations to decidedly local needs (Ostrom, 2010; Copeland and Taylor, 2009; Grafton, 2000).

What are the prospects for this idea to catch on more broadly, given these constraints? To facilitate public participation in polycentric governance arrangements, there is an emerging consensus that stakeholder dialogue requires, at some point, forums that permit face-to-face, structured discussions in which members of the public, government officials, and scientists can clarify sources of technical and political disagreement, and which can guide the decision-making process for which the form of technical solutions has been defined. Melbourne – and to a lesser extent, California and New York City – exemplify progress toward such participation, with Melbourne's Living Melbourne, Living Victoria having made considerably more progress than the other two.

NOTE

1. Special thanks to Nancy Brannon, former doctoral student, for her insights on this topic (N.K. Brannon, "Groundwater: A Community's Management of the Invaluable Resource beneath its Feet", PhD thesis, University of Tennessee, December 2007). Also, for an extended discussion of theories of city growth and water, from which much of this theoretical discussion is taken, see Feldman (2009).

10. Conclusions: some future research needs

Throughout history cities have faced the twin challenges of too much – or too little – water at inopportune times. While these problems are predicted to become even more highly variable due to climate change (as well as varying from city to city), some important generalizations regarding how we might address them are at hand. We have addressed numerous research needs throughout this book, but as a wrap-up, it might be worthwhile to re-focus on some of these initial challenges – the "too much, too little" syndrome, made more urgent by prospects of climate change in this century. These challenges are the subject of this brief chapter.

CLIMATE CHANGE AND WATER: CITY LIFE AT THE EXTREMES

As context, consider that drought and flooding – both symptomatic of extreme weather events – are likely to be exacerbated by climate change. There are anticipated to be five major impacts of climate change upon urban water. The first is saltwater intrusion, likely to be felt most frequently and acutely in cities experiencing sea-level rise. Second is excess heat from roads, buildings and other structures due to urban heat-island effects that can be transferred to stormwater (thereby increasing the temperature of water released into streams, rivers, ponds and lakes). Third, threats to water supply infrastructure due to both drought and flooding generate the need for additional water storage, as well as added measures to fortify sanitation systems that may be already over-taxed (infrastructure). Fourth, in many Third World cities larger numbers of low-income people live on hilly slopes and floodplains especially prone to flooding, storm surges, or other climate-related risks (equity). Finally, climate change threatens future water supplies (and, thus, long-term aspirations).

While in general terms we know and understand the seriousness of these threats and the risks they pose to water infrastructure, much more work needs to be done to investigate their site-specific impacts. More also needs to be done to determine the degree to which low-impact developments and other methods appropriate to the water-sensitive city can meet these threats, and make cities more resilient.

To some degree, New York City, which has been widely studied as a metropolis at the forefront of these threats, affords an example of some of the research challenges facing cities that aspire to become water sustainable in light of extreme weather events. In New York, as in many coastal metropolises, many of the wastewater treatment plants are at elevations of 2–6 m above present sea level and thus within the range of projected surges for tropical storms (hurricanes) and extra-tropical cyclones ("nor'easters") (Rosenzweig and Solecki, 2001).

Moreover, like many U.S. cities along the Atlantic Coast, New York City's vulnerability to storm surges is predominantly from extra-tropical cyclones ("nor'easters") that occur largely between late November and March, and tropical cyclones (hurricanes) that typically strike between July and October. Based on global warming-induced sea-level rise inferred from recent Intergovernmental Panel on Climate Change (IPCC) studies, the recurrence interval for the 100-year flood (probability of occurring in any given year = 1/100) may decrease to 60 years or, under extreme changes, a recurrence interval of as little as four years (Rosenzweig and Solecki, 2001; Jacob et al., 2007).

Increased risk for high sea levels and heavy rains can cause sewer back-up, overflow of water treatment plants. Current activities to address current and future concerns include using sea-level rise forecasts as input to storm surge and elevation models to analyze impact of flooding on NYC coastal facilities. Other concerns include the risk of potential water quality impairment from extreme events and temperature rise due to heavy rains that can increase pathogen levels and turbidity, with possible effects magnified by "first-flush" storms: heavy rains after weeks of dry weather. NYC water supply reservoirs have not been designed for rapid releases and any changes to operations to limit downstream damage through flood control measures will reduce water supply. In addition, adding filtration capacity to the water supply system would be a significant challenge.

Planners in New York have begun to consider these issues by defining risks through probabilistic climate scenarios, and categorizing potential adaptations as related to (1) operations/management; (2) infrastructure; and (3) policy (Rosenzweig et al., 2007). The agency is examining the feasibility of relocating critical control systems to higher floors/ground in low-lying buildings, building protective flood walls, modifying design criteria to reflect changing hydrologic processes, and reconfiguring outfalls to prevent sediment build-up and surging. Several scholars have made the plea that plans for regional capital improvements should be designed to include measures that will reduce vulnerability to the adverse effects of sea-level rise and other threats. Wherever plans are underway for upgrading or constructing new roadways, airport runways, or wastewater treatment plants which may already include flood protection, projected sea-level rise needs to be considered (Beller-Sims et al., 2008).

While considerable thought is being given to strategic decisions and capital investments for water management in response to these varied threats, it remains uncertain to what extent future infrastructure changes can – and will – accommodate room for low-impact developments and other water-sensitive improvements. Moreover, what scale of developments can accommodate future uncertainties? What types of improvements can meet flood risks, water quality and treatment considerations, while also meeting future water supply needs in ways that are adaptable? New York, like many cities, is at the point where it can use this juncture in time to undertake innovative improvements not otherwise impossible. In light of events such as Hurricane Sandy in 2012, perhaps recognition of this window of opportunity is now more acute – a subject for further research.

VULNERABILITY AND ITS VARIED DIMENSIONS

A proper understanding of urban water supply vulnerability must first consider what we term "baseline" sources of water stress caused by urbanization, and then add those distinct stressors that would be generated by climate change.

Little discussed as a research problem in this area of vulnerability are the social justice implications among diverse populations – a subject directly tied to justifications for water-sensitive cities. What are some future research challenges?

Vulnerability and Injustice: Solutions as Problems

Traditionally, environmental justice (EJ) advocates have been concerned with issues such as hazardous and toxic waste storage, disposal, and incineration and their impacts upon racial minorities, the poor, and women. Impacts including land contamination, air- or water-borne pollution, and long-term, intergenerational health hazards that are site-specific and community-centered tend to fall disproportionately upon these groups. In California, this traditional EJ paradigm remains very important for water, particularly in places such as the Central Valley, where low-income communities of color often lack clean water or access to improvements to address toxic contamination (for example, dam-building and flooding ancestral lands; Environmental Justice Coalition, 2005: 10).

By contrast, water conservation, increasing block rate (IBR) pricing, reuse, and desalination do not easily fit into this traditional framework for two reasons. First, the perceived benefits and risks from these innovations are socio-economically cross-cutting. While IBR and metering mostly affect lower-income communities, recycled water and desalination are perceived – fairly or not – as negatively impacting middle class white communities as well as under-represented populations. Second, impacts from all these innovations are perceived as long-term and chronic, rather than short-term and acute (for example, community stigma, diversion of "hard earned tax dollars" to special interests). Thus, they tend not to produce the types of political mobilization associated with hazardous waste or contaminated water supplies – protagonists express concerns through the media or via surveys and polls.

To say that these "newer" conflicts are low-intensity, however, is not to ignore that they represent a growing source of dispute. The traditional and newer idiom for environmental justice and water share a common denominator: the need for fair, open and transparent decision-making processes in which all groups affected by water decisions can equally participate, and where no relevant constituency is excluded. Such processes, we argue, must embrace three characteristics. First, they must be proactive. One cannot wait for public concerns to arise. Decision-makers must reach out to disaffected groups to inform them of the reasons these technologies are being endorsed, to educate and inform them, and to elicit and respond to their concerns. In regard to the three models of environmental justice discussed earlier, we can think of this first characteristic as corresponding to the *covenantal tradition*: everyone is presumed to have a basic right to water and to information about its quality.

Second, these innovations require that attention be paid to compensating those less able to afford the distributional burdens of IBR, appliance retrofits, or even leak repair. While IBR and metering's benefits make good sense, measures are needed to assist special populations in their adoption. Implementation should be calibrated according to affordability. As we have seen, affordability is affected by factors over which the poor may have little control, such as special health needs and care for small children or the elderly. These conditions warrant a kind of *categorical imperative*: counter-balancing for economic hardship with low-income assistance measures, for example, and being more accessible to under-represented groups in scheduling and conducting meetings and accommodating the needs of audiences who lack the technical skill to decipher environmental documents (Environmental Justice Coalition, 2005: 61–3).

Third, potable water innovations like wastewater reuse, graywater use, and stormwater harvesting could benefit from national-level water-supply certification standards that assure protection of in-stream flow and health safeguards. These would affirm that their advantages are independently validated, and strongly resonate with the notion of *stewardship* (Miller, 2006). All three must be embraced to overcome public inhibitions toward adopting these innovations. Further research will be needed to determine what specific societal concerns are likely to arise in specific communities, as well as how these challenges and concerns may be overcome through public engagement.

ADOPTING GREEN INFRASTRUCTURE THROUGH "LEVELING THE PLAYING FIELD"

Throughout this book, we have implicitly juxtaposed low-impact development systems, their characteristics, advantages and disadvantages, and challenges against so-called "conventional" infrastructure. It is now appropriate to ask whether the questions we pose of green infrastructure advocates and their approaches are also applied to conventional infrastructure, and whether we may need to further investigate ways in which the "playing field" for selecting, implementing, and evaluating those approaches which would lead to a water-sensitive city can be made "level" – so to speak.

Three issues are paramount here. First, innovations involving indoor and outdoor water conservation; wastewater reuse; biofilters and other green infrastructure; and, especially, rectifying the impacts and legacy of the urban stream syndrome, will require huge financial investments. Who will pay for these investments? As we have discussed, many cities and water utilities are already strapped paying for repairs and maintenance to existing infrastructures – and we know that conservation rate structures may adversely impact water utilities if declines in water use actually lead to declines in revenues. Will the public support general revenue expenditures for such projects and approaches? In the case of green infrastructure and urban stream syndrome solutions, will private investors be willing to take a risk with new and perhaps untested approaches? What types of cost-sharing approaches and private–public partnerships will be most successful? And, who will bring them about?

Second, green infrastructure and other water-sensitive city innovations may take a long time to show positive results as exemplified by improvement in the quality of receiving waters, more efficient and sustainable water uses, and improvements in the local built environment and its economic valuation, as well as aesthetics. Can decision-makers and the public be patient enough to "wait out" these results? Can we avoid the "immediate gratification" syndrome so prevalent in modern day public policies generally? And, can the benefits of a set of approaches that might not yield positive benefits until far into the future – benefits that are realized by our grandchildren and great-grandchildren – be sufficient to motivate policymakers to introduce these approaches?

Earlier, we suggested that the basic underlying advantages of a water-sensitive city lie in its capacity to meld two important and increasingly contradictory goals: achieving resilience – the capacity of an urban water system to undergo shocks from climate change or other environmental stressors while retaining essentially the same function, structure, feedbacks, and identity as before (see Kiparsky et al., 2013; Dobbie et al., 2014; Brown et al., 2009) – and adaptiveness – the capacity to experience "adaptive transitions" in order to positively adjust to external environmental or other pressures in unplanned and improvisational ways.

Third, while resilience implies a top-down, centralized approach to coping with the pressures of climate change, population growth and pollution, adaptiveness suggests the possibility of a "bottom-up" and decentralized capacity to nimbly and democratically evolve. In actuality, few cities have actually achieved this – at least in its entirety (Ferguson et al., 2013; Kiparsky et al., 2013; Gleick, 2003). As we saw, changes required to bring this about include not only rethinking "hard" engineering approaches but changes in law, institutions, market incentives, and educational programs – including social marketing. More research on these problems is needed, as is collaboration among social and policy scientists, urban planners, engineers, and physical and life science experts.

TOWARD A WATER-SENSITIVE CITY: THE CHALLENGES AHEAD

As we have seen in our discussion of the urban stream syndrome and other problems, there are growing demands in cities especially to treat freshwater as an amenity as opposed to purely an economic commodity. There also are growing demands to rethink how we value, allocate, and prioritize its use, particularly in light of major drought and serious apprehensions about the long-term impacts of climate change.

Both sets of demands have become linked in greater calls to manage freshwater more *equitably* – a word that has multiple, and even conflicting meanings: fairness toward people and their welfare, or toward nature and its sustainability. In cases where certain groups seek to develop large dams or diversion projects, for instance, there may be fervent pressure to ensure that underrepresented groups living adjacent to these proposed projects share in their benefits. To others, however, equity with respect to building dams may mean protecting fauna and flora that might be driven to extinction if such projects are built without regard to their ecological impact.

Typically, political decisions over water supply in Third World megacities tend to be made by engineers and planners working for local water utilities, and also by external agencies (for example, the World Bank) that fund efforts to improve delivery and treatment infrastructure (Bakker, 2013). Decision-making power is, thus, concentrated in the hands of an expert few (see Chapter 1), and long-standing decision-making processes constrain the selection of policy alternatives by favoring wealthier as opposed to disadvantaged local residents.

In many cities in the developing world, water distribution networks are frequently based on traditional urban planning patterns initially planned by colonial powers to favor delivery to wealthier districts as opposed to poorer ones. Often, high-quality piped water was made available to the affluent, and polluted surface water was the only option available to the poor (Bakker, 2013). A good example is Mumbai, India's economic powerhouse (historically called Bombay under the British Raj). While Bombay was the leading manufacturing and commercial center of India, and supported a large and diverse population of workers, development of an urban water infrastructure, including proper drainage and sewage service, was late in coming and only introduced after the pleas of public hygiene reformers such as Florence Nightingale. Typically, British and other European colonists neglected the introduction of modern sanitary sewers, waste disposal systems, and reliable sources of water supply (Hunt, 2015; Sule, 2003).

While progress has been made in improving water provisioning and sanitation in the world's cities – measured statistically – the need to upgrade these services in light of continued population growth constitutes a race against time. Over the next few decades, the world's urban population is expected to increase from some 4 billion to 6.3 billion by 2050 – a higher rate of growth (36 percent) than the overall projected growth of the world's population (26 percent) during this same period (UN Water, 2014).

While innovations exist that can be applied to these cities, questions remain regarding their benefit, cost, and decision-making. In China, for instance, rainwater harvesting has taken on a special importance in urban plans in Beijing and other cities because it not only conserves water but abates flooding, groundwater depletion, and rainwater runoff pollution (Wong, 2007). In essence, for urban water management, rainwater harvesting can help address a number of water quality as well as water supply needs, as well as help attenuate the effects from climate variability – and it is a relatively low-cost tool for addressing China's water problems.

Unanswered political questions remain, however, applicable elsewhere.

Will the price of water charged to consumers be calibrated in such a way as to encourage conservation as well as adoption of rainwater harvesting, especially for easily substitutable non-potable uses? Will decisions regarding its adoption be made in a thoughtful manner to ensure that some harvesting "solutions" do not exacerbate water problems in the future? And, is the process through which innovation decisions are made subject to broad public and policymaker discussion? As we have seen throughout this book – these are ubiquitous challenges for the water-sensitive city.

Appendix: the legal regime for water-sensitive cities

THE INSTITUTIONAL, MANAGEMENT AND POLICY APPROACHES IN CALIFORNIA AND AUSTRALIA

California

AB 325: Water Conservation in Landscape Act (1990)

- Requires DWR to convene advisory task force and create a model landscape ordinance.
- Model Ordinance for landscape adopted in 1992.
- One element of the ordinance: landscape water budgets.
- Maximum Applied Water Allowance (MAWA) established based on landscape area and climate where landscape is located.

AB 2717: California Urban Water Conservation Council (2004)

- Directed CUWCC to convene a task force of public agencies, private agencies, and landscape experts to develop recommendations/ proposals to improve CA's water use efficiency.
- Stakeholder task force adopted 43 recommendations (including updating AB 325).
- Many recommendations suggest updating the Model Ordinance.

AB 1881: Water Efficient Landscape Ordinance (2006)

- Requires DWR to update the Model Ordinance reflecting recommendations from 2004 report: Water Smart Landscapes for California.
- By January 2010, local water agencies must adopt or a similar ordinance with water-conserving potential.
- In service areas of local agencies that fail to pass an ordinance, the Model Ordinance will be applicable (even in charter cities).

SBX7-7: Water Conservation Ordinance (2009)
http://www.water.ca.gov/wateruseefficiency/sb7/

- Urban water agencies must reduce per capita water consumption in their service area 10 percent by 2015 and 20 percent by 2020.
- Must include baseline daily per capita water use, water use target, interim water use target, and compliance daily per capita water use in 2010 UWMP update.
- Effective 2016, urban retail water agencies who do not meet water conservation requirements established by this Bill will not be eligible for state water grants or loans.

AB 1750: Rainwater Capture Act (2012)
http://www.ncsl.org/research/environment-and-natural-resources/rain
water-harvesting.aspx
http://leginfo.legislature.ca.gov/faces/billNavClient.xhtml?bill_id=201120
120AB1750

- Would enact the Rainwater Capture Act of 2012. Would authorize residential, commercial and governmental landowners to install, maintain, and operate rain barrel systems and rainwater capture systems for specified purposes, provided that the systems comply with specified requirements.
- Would authorize a landscape contractor working within the classification of his or her license to enter into a prime contract for the construction of a rainwater capture system if the system is used exclusively for landscape irrigation.
- Now legal for residential users and private and public entities to capture and reuse rainwater as an onsite water supply and to recharge groundwater.
- Departs from long-standing legal tradition making it illegal to capture and use precipitation based on the prior appropriation doctrine.

The Act exempts the capture and use of rainwater from rooftops from the State Water Resources Control Board's (SWRCB) permitting authority over appropriations of water. http://www.lexisnexis.com/legalnewsroom/ top-emerging-trends/b/emerging-trends-law-blog/archive/2013/02/04/ california-s-rainwater-recapture-act-lets-state-residents-capture-use-harvested-rainwater.aspx.

Prior to enactment of the Act, the SWRCB required all would-be appropriators to apply for and obtain a permit to appropriate water from any source, including water falling in the form of precipitation. Under the Act, however, the use of rainwater – defined as "precipitation on any public or private parcel that has not entered an offsite storm drain system or channel, a flood channel, or any other stream channel, and has not been previously been put to beneficial use" – is not subject to the California Water Code's SWRCB permit requirement (California Water Code §§ 1200 *et seq.*). Relief from the permit requirement enables residents, private businesses and public agencies to create new onsite water supplies to meet landscaping needs, thus decreasing the use of potable water to meet those needs. The language of the Act recognizes that it may contribute to attainment of the statewide "20 × 2020 goal," which aims to achieve a 20 percent reduction in urban per capita potable water demand by 31 December 2020 (2012 Cal. Stats. ch. 537, Sec. 2).

Proposition 218

- Constitutional amendment that requires local units of government to seek approval of a simple majority of ratepayers or a two-thirds majority of landowners before imposing any new fees or taxes.
- Requires that all taxes and fees imposed by a public agency must not be any greater than the "cost of service" to the individual.
- Affects tiered rate structures: *Capistrano Taxpayers Association* v. *City of San Juan Capistrano* (2013).

Lessons from Capistrano:

Option 1: Public agencies must work with their rate consultants to establish a clear, understandable and compelling record for the courts that demonstrate that new rates will result in water savings, that new rates are equitable, and that higher water uses place a disproportionate burden on water systems, particularly in the cost of new supplies, and therefore warrant higher rates.

Option 2: Structure higher tiers as penalties, which are exempted from Prop. 218. Characterizing excessive water use as wasteful and subsequently penalizing those actions would appear to be a reasonable action, although likely to meet with political opposition from some.

Proposition 26 (2010): Local Agency Guidelines for Compliance
http://www.acwa.com/sites/default/files/post/state-budget-fees/2012/07/
prop-26.pdf

- Constitutional amendment that broadly redefines the term "tax," which is subject to voter approval.
- Local water agencies seeking to adopt, increase or extend any fee or charge now bear the burden of proving that the fee or charge is not a tax.
- Agencies also must be prepared to document that the amount of the fee or charge is no more than necessary to cover the reasonable costs of the activity or service being provided, and that the manner in which those costs are allocated to a payer bears a fair or reasonable relationship to the payer's burdens on, or benefits received from, the activity or service.

SCRCB Recycled Water Policy (2009)
http://socalwater.org/images/SCWC_Stormwater_White_Paper__Case_
Studies.Smaller.pdf

- The State Water Board's Recycled Water Policy adopted in 2009 encourages "local and regional water agencies to move toward clean, abundant, local water for California by emphasizing appropriate water recycling, water conservation, and maintenance of supply infrastructure and the use of stormwater (including dry-weather urban runoff)."
- Towards that end, one of the measurable goals adopted in the policy is for California to increase the use of stormwater from the 2007 baseline by at least 500 000 acre-feet per year by 2020 and by at least one million acre-feet per year by 2030. As for the State Water Board's part in this effort, the preamble of the policy notes that the "State Water Board expects to develop additional policies to encourage the use of stormwater, encourage water conservation, encourage the conjunctive use of surface and groundwater, and improve the use of local water supplies."
- The policy also recognizes the need for regulatory flexibility in this collaborative, statewide effort to achieve sustainable local water supply: "The State Water Board also encourages the Regional Water Boards to require less stringent monitoring and regulatory requirements for stormwater treatment and use projects than for projects involving untreated stormwater discharges."

SWCRB Water Conservation Fine (2014)
http://www.swrcb.ca.gov/press_room/press_releases/2014/pr071514.pdf

- Intended to reduce urban outdoor water use.
- Larger water suppliers will be required to activate their Water Shortage Contingency Plan to a level where outdoor irrigation restrictions are mandatory.
- In communities where no water shortage contingency plan exists, the regulation requires that water suppliers either limit outdoor irrigation to twice a week or implement other comparable conservation actions.
- Large water suppliers must report water use on a monthly basis to track progress.
- Local agencies could ask courts to fine water users up to $500 a day for failure to implement conservation requirements in addition to their existing authorities and processes.
- The State Water Board could initiate enforcement actions against water agencies that don't comply with the new regulations. Failure to comply with a State Water Board enforcement order by water agencies is subject to up to a $10 000 a day penalty.

Melbourne, Victoria and Australia

Office of Living Victoria: implements Integrated Water Management Policy
http://www.depi.vic.gov.au/water/saving-water/saving-water-at-home/using-greywater

Graywater

- Laundry to landscape systems do not require a permit if laundry water is immediately used on the landscape; includes clothes washer, baths, showers, and sinks and excludes kitchen sink and dishwasher.
- Permanent graywater systems need to be EPA approved systems and require a permit for operation from local municipality.

Water conservation hierarchy and inherent risk of alternative water resources
http://www.epa.vic.gov.au/your-environment/water/reusing-and-recycling-water#graywater

- Approach is to avoid excessive water consumption, wastewater, and/or runoff.

Permanent water savings rules
http://www.depi.vic.gov.au/water/saving-water/water-restrictions/permane
nt-water-saving-rules

- Water saving rules are in effect at all times and there are penalties for violating the rules.
- Individuals may apply for exemptions or variances.
- Based on the Uniform Drought Water Restrictions Guidelines http://www.depi.vic.gov.au/water/saving-water/water-restrictions.

Bibliography

Adekalu, K.O., J.A. Osunbitan and O.E. Ojo (2002). "Water sources and demand in South Western Nigeria: implications for water development planners and scientists," *Technovation*, **22**, 799–805.

AghaKouchak, A., D. Feldman, M.J. Stewardson, J.D. Saphores, S. Grant and B. Sanders (2014). "Australia's drought: lessons for California," *Science*, **343** (6178), 1430–31.

Ahiablame, L.M., B.A. Engel and I. Chaubey (2012). "Effectiveness of low impact development practices: literature review and suggestions for future research," *Water, Air, & Soil Pollution*, **223** (7), 4253–73.

Ahmed, W., T. Gardner and S. Toze (2011). "Microbial quality of roof-harvested rainwater and health risks: a review," *Journal of Environmental Quality*, **40**, 13–21.

Akbari, H. (2002). "Shade trees reduce building energy use and CO_2 emissions from power plants," *Environmental Pollution*, **116**, S119–S126.

Allen, S. (2012). "Water rate hike prompts inquiry," *Los Angeles Times*, 4 March, A33.

American Museum of Natural History (2011). "The New York water story," available at http://www.amnh.org/education/resources/rfl/web/nycwater/AMNH_Water.php.

American Planning Association (2011). "How cities use parks to improve public health," City Parks Forum Briefing Paper #7, available at http://www.planning.org/cityparks/briefingpapers/physicalactivity.htm.

Anand, C. and D.S. Apul (2011). "Economic and environmental analysis of standard, high efficiency, rainwater flushed, and composting toilets," *Journal of Environmental Management*, **92** (3), 419–28.

Aquacraft (2011). "California single family water use efficiency study," prepared in coordination with the Irvine Ranch Water District, Stratus Consulting, The Pacific Institute, and the California Department of Water Resources, June, available at http://www.irwd.com/images/pdf/save-water/CaSingleFamilyWaterUseEfficiencyStudyJune2011.pdf.

Araral, Ed and Kris Hartley (2013). "Polycentric governance for a new environmental regime: theoretical frontiers in policy reform and public administration," paper presented at the international Conference on Public Policy, June, Sciences Po, Grenoble.

Arbues, F., M.Á. García-Valiñas and R. Martinez Espiñeira (2003). "Estimation of residential water demand: a state-of-the-art review," *The Journal of Socio-Economics*, **32**, 81–102.

Askarizadeh, A., M.A. Rippy, T.D. Fletcher, D.L. Feldman, J. Peng, P. Bowler, A.S. Mehring et al. (2015). "From rain tanks to catchments: use of low-impact development to address hydrologic symptoms of the urban stream syndrome," *Environmental Science and Technology*, **49** (19), 11264–80.

Australian Government (2004). "National water initiative," Canberra: Department of the Environment.

Australian National Audit Office (2009). "Innovation in the Public Sector: Enabling Better Performance, Driving New Directions. Better Practice Guide," December, available at http://www.anao.gov.au/uploads/docu ments/Innovation_in%20the_Public_Sector.pdf.

AWWA (American Water Works Association) (2012). "Manual of water supply practices, m6. water meters – selection, installation, testing, and maintenance," 5th edn, Denver.

Baer, K.E. and C.M. Pringle (2000). "Special problems of urban river conservation: the encroaching megalopolis," in P.J. Boon, B.R. Davies and G.E. Potts (eds), *Global Perspectives on River Conservation: Science, Policy, and Practice*, New York: John Wiley, pp. 385–402.

Baer, M. (2008). "The global water crisis, privatization, and the Bolivian water war," in J.M. Whitely, H. Ingram and R.W. Perry (eds), *Water, Place, and Equity*, Cambridge, MA: MIT Press, pp. 195–224.

Bakker, K. (2013). "Constructing 'public' water: the World Bank, urban water supply, and the biopolitics of development," *Environment and Planning D: Society and Space*, **31**, 280–300.

Barker, F., R. Faggian and A.J. Hamilton (2011). "A history of wastewater irrigation in Melbourne, Australia," *Journal of Water Sustainability*, **2**, 31–50.

Barlow, Maude and Tony Clarke (2002). *Blue Gold: The Fight to Stop the Corporate Theft of the World's Water*, New York: The New Press.

Baumann, D.D., J.J. Boland and Michael W. Hanemann (1997). *Urban Water Demand Management and Planning*, New York: McGraw-Hill.

Bedan, E.S. and J.C. Clausen (2009). "Stormwater runoff quality and quantity from traditional and low impact development water-sheds," *Journal of the American Water Resources Association*, **45** (4), 998–1008.

Beller-Simms, Nancy, Helen Ingram, David Feldman, Nathan Mantua and Katharine L. Jacobs (2008). "U.S. climate change science program synthesis and assessment product 5.3 – decision-support experiments and evaluations using seasonal to interannual forecasts and observational data: a focus on water resources," National Oceanic and Atmospheric Administration, November, available at http://www.climatescience.gov/Library/sap/sap5-3/final-report/#finalreport.

Berndtsson, J.C. (2010). "Green roof performance towards management of runoff water quantity and quality: a review," *Ecological Engineering*, **36** (4), 351–60.

Bhaskar, A.S. and C. Welty (2015). "Analysis of subsurface storage and streamflow generation in urban watersheds," *Water Resources Research*, **51** (3), 1493–513.

Biggs, C., C. Ryan and J. Wiseman (2010a). "Localised solution: building capacity and resilience with distributed production systems," Victorian Eco-Innovation Lab, University of Melbourne.

Biggs, C., C. Ryan and J. Wiseman (2010b). "Distributed systems: a design model for sustainable and resilient infrastructure," Victorian Eco-Innovation Lab, University of Melbourne.

Biggs, C., C. Ryan, J. Wiseman and K. Larsen (2009). "Distributed water systems: a networked and localized approach for sustainable water services," Victorian Eco-Innovation Lab, University of Melbourne.

Bijoor, N.S., C.I. Czimczik, D.E. Pataki and S.A. Billings (2008). "Effects of temperature and fertilization on nitrogen cycling and community composition of an urban lawn," *Global Change Biology*, **14**, 2119–31.

Bijoor, N.S., H.R. McCarthy, D. Zhang and D.E. Pataki (2012). "Water sources of urban trees in the Los Angeles metropolitan area," *Urban Ecosystems*, **15**, 195–214.

Bijoor, N., D. Pataki, D. Haver and J. Famiglietti (2014). "A comparative study of the water budgets of lawns under three management scenarios," *Urban Ecosystems*, **17** (4), 1095–117.

Black & Veatch (1995). *California Water Charge Survey*, Los Angeles: Black & Veatch.

Black & Veatch (2006). *California Water Rate Survey*, Los Angeles: Black & Veatch.

Boatright, Mary T., Daniel J. Gargola and Richard A. Talbert (2004). *The Romans: From Village to Empire – A History of Ancient Rome from Earliest Times to Constantine*, Oxford: Oxford University Press.

Boberg, J. (2005). "Liquid assets: how demographic changes and water management policies affect freshwater resources," Report-MG-358-CF, Santa Monica, CA: RAND Corporation.

Booth, D.B. (1991). "Urbanization and the natural drainage system: impacts, solutions, and prognoses," *Northwest Environmental Journal*, **7** (1), 93–118.

Booth, D.B. and C.R. Jackson (1997). "Urbanization of aquatic systems: degradation thresholds, storm water detection, and the limits of mitigation," *Journal of the American Water Resources Association*, **33** (5), 1077–90.

Bos, D.G. and H.L. Brown (2015). "Overcoming barriers to community participation in a catchment-scale experiment: building trust and changing behavior," *Freshwater Science*, **34** (3), 1169–75.

Braemer, F., B. Geyer, C. Castel and M. Abdulkarim (2010). "Conquest of new lands and water systems in the western Fertile Crescent (Central and Southern Syria)," *Water History*, **2**, 91–114.

Brattebo, B.O. and D.B. Booth (2003). "Long-term storm water quantity and quality performance of permeable pavement systems," *Water Research*, **37** (18), 4369–76.

Brooks, D.B. and O.M. Brandes (2011). "Why a water soft path, why now and what then?," *International Journal of Water Resources Development*, **27**, 315–44.

Brooks, D.B., O.M. Brandes and S. Gurman (2009). *Making the Most of the Water we have: The Soft Path Approach to Water Management*, London: Earthscan.

Brown, A.E., L. Zhang, T.A. McMahon, A.W. Western and R.A. Vertessy (2005). "A review of paired catchment studies for determining changes in water yield resulting from alterations in vegetation," *Journal of Hydrology*, **310** (1), 28–61.

Brown, R.R. and M.A. Farrelly (2009). "Delivering sustainable urban water management: a review of the hurdles we face," *Water Science & Technology*, **59** (5), 839–46.

Brown, R., M. Farrelly and D. Loorbach (2013). "Actors working the institutions in sustainability transitions: the case of Melbourne's stormwater management," *Global Environmental Change*, **23**, 701–18.

Brown, R.R. and N.A. Keath (2008). "Drawing on social theory for transitioning to sustainable urban water management: turning the institutional super-tanker," *Australian Journal of Water Resources*, **12** (2), 73–83.

Brown, R., N. Keath and T. Wong (2008). "Transitioning to water sensitive cities: historical, current and future transition states," paper presented at the 11th International Conference on Urban Drainage, Edinburgh.

Brown, R.R., N. Keath and T.H.F. Wong (2009). "Urban water management in cities: historical, current, and future regimes," *Water Science & Technology*, **59** (5), 847–55.

Bryner, Gary and Elizabeth Purcell (2003). *Groundwater Law Sourcebook of the Western United States*, Boulder, CO: Natural Resources Law Center, University of Colorado Law School.

Bunn, S.E. and A.H. Arthington (2002). "Basic principles and ecological consequences of altered flow regimes for aquatic biodiversity," *Environmental Management*, **30** (4), 492–507.

Burns, M.J., T.D. Fletcher, C.J. Walsh, A.R. Ladson and B.E. Hatt (2012). "Hydrologic shortcomings of conventional urban storm water management and opportunities for reform," *Landscape Urban Plan*, **105** (3), 230–40.

Burns, M.J., T.D. Fletcher, C.J. Walsh, A.R. Ladson and B.E. Hatt (2013). "Setting objectives for hydrologic restoration: from site-scale to catchment-scale," paper presented at *NOVATECH 2013*, Lyon, 23–27 June.

Burns, M.J., E. Wallis and V. Matic (2015). "Building capacity in low-impact drainage management through research collaboration," *Freshwater Science*, **34** (3), 1176–85.

Burton, Lloyd (1991). *American Indian Water Rights and the Limits of Law*, Lawrence, KS: University Press of Kansas.

Buttle, J.M. (1994). "Isotope hydrograph separations and rapid delivery of pre-event water from drainage basins," *Progress in Physical Geography*, **18** (1), 16–41.

Cahill, Ryan and Jay Lund (2011). "Residential water conservation in Australia and California," Working Paper, Department of Civil and Environmental Engineering, University of California, Davis, November.

California Department of Water Resources (2013a). "A commitment to action: perspectives from California's first integrated water management summit," *Water 360*, April, available at http://www.water.ca.gov/irwm/other_resources/pdfs/water360proceedings.pdf.

California Department of Water Resources (2013b). "California Water Plan update 2013: public draft review," available at http://www.water plan.water.ca.gov/cwpu2013/prd/index.cfm.

California Department of Water Resources (2014). "About DWR: Integrated Water Management," available at http://www.water.ca.gov/about/regional.cfm.

California Energy Commission (2006). "Refining estimates of water-related energy use in California," Report CEC-500-2006-118, California Energy Commission, Sacramento, CA.

California Senate Bill 7 (2009). "The Water Conservation Act of 2009," also known as "California Senate Bill 7," available at http://www.water.ca.gov/wateruseefficiency/sb7/.

California State Water Resources Control Board (2014). "Storm water – municipal permits," available at http://www.waterboards.ca.gov/losangeles/water_issues/programs/stormwater/municipal/.

Cardenas-Lailhacar, B. and M.D. Dukes (2012). "Soil moisture sensor landscape irrigation controllers: a review of multi-study results and future implications," *Transactions of the Asabe*, **55**, 581–90.

Center for Demographic Research (2009). "Water usage in Orange County: 2007–2008," *Orange County Profiles*, **14** (1), 1–4.

Charlesworth, S.M., E. Harker and S. Rickard (2003). "A review of sustainable drainage systems (SuDS): a soft option for hard drainage questions?," *Geography*, **88** (2), 99–107.

Chestnutt, T. and C. McSpadden (1991). *A Model-based Evaluation of Westchester Water Conservation Programs*, San Diego: A & N Technical Services.

City of Los Angeles (2008). "Securing L.A.'s water supply," Los Angeles, CA: Department of Water and Power.

City of Los Angeles (2011). *Development Best Management Practices Handbook: Low Impact Development Manual – Part B Planning Activities*, 4th edn, June.

City of Los Angeles (2014). "LA Stormwater: LA's Watershed Protection Program," available at http://www.lastormwater.org/.

City of Melbourne (2014). "Amendment C142: stormwater management (water sensitive urban design)," available at http://www.melbourne.vic.gov. au/BuildingandPlanning/Planning/planningschemeamendments/Pages/ AmendmentC142.aspx.

Clayton, J.A. (2009). "Market-driven solutions to economic, environmental, and social issues related to water management in the western USA," *Water*, **1**, 19–31.

Coarelli, Filippo (2007). *Rome and Environs: An Archaeological Guide*, Berkeley, CA: University of California Press.

Connolly, Peter and Hazel Dodge (1998). *The Ancient City: Life in Classical Athens and Rome*, Oxford: Oxford University Press.

Coombes, P.J. and G. Kuczera (2003). "Analysis of the performance of rainwater tanks in Australian capital cities," in *Proceedings of 28th International Hydrology and Water Resources Symposium*, Wollongong, NSW, 10–14 November, pp. 235–42.

Copeland, Brian R. and M. Scott Taylor (2009). "Trade, tragedy, and the commons," *American Economic Review*, **99** (3), 725–49.

Corbett, D. (2010). "Achieving sustainable stormwater management in Melbourne, Australia, as part of the journey to a water sensitive city," paper presented at *NOVATECH 2010*, available at http://docu ments.irevues.inist.fr/bitstream/handle/2042/35638/13101-015COR.pdf? sequence=1.

Coulomb, Rene (2001). "Speech presented by the Vice President of the World Water Council at the Closing Session of the 11th Stockholm Water Symposium," World Water Council – 3rd World Water Forum – Stockholm Water Symposium, 16 August, available at www.worldwater council.org.

Coutts, A.M., E. Daly, J. Beringer and N.J Tapper (2013). "Assessing practical measures to reduce urban heat: green and cool roofs," *Building and Environment*, **70**, 266–76.

Crase, L. (2008). *Water Policy in Australia: The Impact of Change and Uncertainty*, Washington, DC: RFF Press.

Crase, L. (2009). "Dynamic community preferences: lessons for institutional design and measuring transaction and transformation costs," in L. Crase and V.P. Gandhi (eds), *Reforming Institutions in Water Resource Management: Policy and Performance for Sustainable Development*, London: Earthscan, pp. 45–61.

Cronon, W. (1992). *Nature's Metropolis: Chicago and the Great West*, New York: W.W. Norton.

Croton Watershed Clean Water Coalition (2009). "Updated 2009 Croton Watershed Management Plan," New York: CWCWC, available at http://www.newyorkwater.org/pdf/managementPlan/MPlanNOV309.pdf.

Crow, Ben and Sultana Farhana (2002). "Gender, class, and access to water: three cases in a poor and crowded delta," *Society and Natural Resources*, **15** (8), 709–24.

CSIRO (1999). "Water sensitive urban design," in *Urban Stormwater: Best Practice Environmental Management Guidelines*, CSIRO Publishing, pp. 47–62.

Cuffney, T.F., R.A. Brightbill, J.T. May and I.R. Waite (2010). "Responses of benthic macroinvertebrates to environmental changes associated with urbanization in nine metropolitan areas," *Ecological Applications*, **20** (5), 1384–401.

Cuffney, T.F. and J.F. Falcone (2008). *Derivation of Nationally Consistent Indices Representing Urban Intensity within and across Nine Metropolitan Areas of the Conterminous United States*, U.S. Geological Survey Scientific Investigations Report 2008-5095; Washington, DC: U.S. Geological Survey.

Culture Victoria (2016). "Melbourne and Smellbourne," Public Record Office of Victoria: Culture Victoria, available at http://www.cv.vic.gov.au/stories/built-environment/melbourne-and-smellbourne/.

Daigger, G.T. (2009). "Evolving urban water and residuals management paradigm: water reclamation and reuse, decentralization, and resource recovery," *Water Environment Research*, **81** (8), 809–23.

Daigger, G.T. (2011). *Sustainable Urban Water and Resource Management*, Washington, DC: National Academy of Engineering.

Dalhuisen, J.M., R. Florax, H.L.F. de Groot and P. Nijkamp (2003). "Price and income elasticities of residential water demand: a meta-analysis," *Land Economics*, **79**, 292–308.

Davis, C. and S. Slater (2013). "California's Rainwater Recapture

Act lets state residents capture, use harvested rainwater," JDSupra Business Advisor, available at http://www.jdsupra.com/legalnews/californias-rainwater-recapture-act-let-66504/.

Davis, Margaret Leslie (1993). *Rivers in the Desert: William Mulholland and the Inventing of Los Angeles*, New York: HarperCollins.

Derthick, Martha (1974). *Between State and Nation: Regional Organizations of the United States*, Washington, DC: Brookings Institution.

Deverell, W. and G. Hise (2005). *Land of Sunshine: An Environmental History of Metropolitan Los Angeles*, Pittsburgh, PA: University of Pittsburgh.

Dobbie, M.F., K.L. Brookes and R.R. Brown (2014). "Transition to water-cycle city: risk perceptions and receptivity of Australian urban water practitioners," *Urban Water Journal*, **11** (5), 427–43.

Doig, Will (2012). "The impending urban water crisis," *Salon.com*, available at http://www.salon.com/2012/.

Dolnicar, S. and A. Hurlimann (2010). "Acceptance of water alternatives in Australia – 2009," *Water Science & Technology*, **61** (8), 2137–42.

Dolnicar, S., A. Hurlimann and B. Grun (2012). "Water conservation behavior in Australia," *Journal of Environmental Management*, **105** (114), 44–52.

Dolnicar, S., A. Hurlimann and L. Nghiem (2010). "The effect of information on public acceptance: the case of water from alternative sources," *Journal of Environmental Management*, **91** (6), 1288–93.

Donnelly, Kristina and Juliet Christian-Smith (2013). "California water rates and the 'new normal'," Oakland, CA: Pacific Institute.

Doolan, Jane (2015). "Lessons from Australia's Millennium Drought," Canberra: University of Canberra and National Water Commission, unpublished presentation, Public Policy Institute of California, San Francisco.

Dowling, J. (2013). "Is the wally back? Melbourne water use surges?," *The Age Victoria*, 18 January, available at http://www.theage.com.au/victoria/is-the-wally-back-melbourne-water-use-surges-20130117-2cwan.html.

Downs, T.J., M. Mazari-Hiriart, R. Domínguez-Mora and I.H. Suffet (2000). "Sustainability of least cost policies for meeting Mexico City's future water demand," *Water Resources Research*, **36** (8), 2321–39.

Du Pisani, P.L. (2006). "Direct reclamation of potable water at Windhoek's Goreangab reclamation plant," *Desalination*, **188** (1–3), 79–88.

Dukes, M.D. (2012). "Water conservation potential of landscape irrigation smart controllers," *Transactions of the ASABE 55*, pp. 563–9.

Eades, Mark (2012). "Residents criticize water rate increase," *Orange County Register*, 4 March, p. 3.

Elcock, D. (2009). "Baseline and projected water demand data for energy and competing water use sectors," U.S. Department of Energy, ANL/ EUS/TM/08-8 for US DOE/NETL.

Emerson, C.H., C. Welty and R.G. Traver (2005). "Watershed-scale evaluation of a system of storm water detention basins," *Journal of Hydrologic Engineering*, **10** (3), 237–42.

Endreny, T. (2008). "Naturalizing urban watershed hydrology to mitigate urban heat-island effects," *Hydrological Processes*, **22** (3), 461–3.

Engle, N.L., O.R. Johns, M.C. Lemos and D.R. Nelson (2011). "Integrated and adaptive management of water resources: tensions, legacies, and the next best thing," *Ecology & Society*, **16** (1).

Environmental Justice Coalition for Water (2005). "Thirsty for justice: a people's blueprint for California," Oakland, CA: Environmental Justice Coalition for Water.

Environment Australia (1994). "The Council of Australian Governments' Water Reform Framework: extracts from Council of Australian Governments," Hobart, 25 February, Communiqué, Canberra: Environment Australia, available at http://webarchive.nla.gov.au/gov/ 20130904083606/, http://www.environment.gov.au/water/publications/ action/policyframework.html.

Environment Australia (2002). "Introduction to urban stormwater management in Australia," prepared under the Urban Stormwater Initiative of the Living Cities Program 2002, Canberra, Australia.

Erie, Steven P. (2006). *Beyond Chinatown: The Metropolitan Water District, Growth, and the Environment in Southern California*, Palo Alto, CA: Stanford University Press.

Espey, M., J. Espey and W.D. Shaw (1997). "Price elasticity of residential demand for water: a meta-analysis," *Water Resources Research*, **33**, 1369–74.

Facility for Advancing Water Biofiltration (FAWB) (2009). "Guidelines for filter media in biofiltration systems," Version 3.01, Sydney: Facility for Advancing Water Biofiltration (FAWB).

Feldman, David L. (2009). "Preventing the repetition: or, what Los Angeles' experience in water management can teach Atlanta about urban water disputes," *Water Resources Research*, **45** (4).

Ferguson, B.C., R.R. Brown, N. Frantzeskaki, F.J. de Haan and A. Deletic (2013). "The enabling institutional context for integrated water management: lessons from Melbourne," *Water Research*, **47** (20), 7300–14.

Ferguson, B.C., R.R. Brown, F.J. de Haan and A. Deletic (2014). "Analysis of institutional work on innovation trajectories in water infrastructure systems of Melbourne, Australia," *Environmental Innovation and Societal Transitions*, **15**, 42–64.

Fletcher, T.D., H. Andrieu and P. Hamel (2013a). "Understanding, management and modeling of urban hydrology and its consequences for receiving waters: a state of the art," *Advances in Water Resources*, **51**, 261–79.

Fletcher, T.D., A. Deletic, V.G. Mitchell and B.E. Hatt (2008). "Reuse of urban runoff in Australia: a review of recent advances and remaining challenges," *Journal of Environmental Quality*, **37** (5 Suppl.), S-116.

Fletcher, T.D., W. Shuster, W.F. Hunt, R. Ashley, D. Butler, S. Arthur, S. Trowsdale et al. (2013b). "SUDS, LID, BMPs, WSUD and more: the evolution and application of terminology surrounding urban drainage," *Urban Water Journal*, **12** (7), 525–42.

Fogelson, Robert M. (1993). *The Fragmented Metropolis: Los Angeles, 1850–1930*, Berkeley and Los Angeles, CA: University of California Press.

Freeman, Charles (2004). *Egypt, Greece, and Rome: Civilizations of the Ancient Mediterranean*, 2nd edn, Oxford: Oxford University Press.

Frost, Lionel (2013). "A research agenda for understanding urban water history," unpublished talk given by Professor Lionel Frost, Monash University, July.

Fuenfschilling, L. and B. Truffer (2014). "The structuration of socio-technical regimes: conceptual foundations from institutional theory," *Research Policy*, **43**, 772–91.

Gandy, M. (2008). "Landscapes of disaster: water, modernity, and urban fragmentation in Mumbai," *Environment and Planning*, **40**, 108–30.

Garnsey, Peter and Richard Saller (1987). *The Roman Empire: Economy, Society, and Culture*, Berkeley, CA: University of California Press.

Garrison, N., C. Kloss, R. Lukes and J. Devine (2011). "Capturing rainwater from rooftops: an efficient water resource management strategy that increases supply and reduces pollution," Washington, DC: Natural Resources Defense Council.

Geels, F.W. (2006). "The hygienic transition from cesspools to sewer systems (1840–1930): the dynamics of regime transformation," *Research Policy*, **35**, 1069–82.

Geels, F.W. and J. Schot (2007). "Typology of sociotechnical transition pathways," *Research Policy*, **36**, 399–417.

Geo-Mexico (2013). "Mexico's major cities confront serious water supply issues," available at http://geo-mexico.com/?p=9034.

Gersonius, B., R. Ashley, A. Pathirana and C. Zevenbergen (2013). "Climate change uncertainty: building flexibility into water and flood risk infrastructure," *Climate Change*, **116**, 411–23.

Giacomoni, M.H., E.M. Zechman and K. Brumbelow (2012). "Hydrologic footprint residence: environmentally friendly criteria for best management practices," *Journal of Hydrologic Engineering*, **17** (1), 99–108.

Giddens, A. (1984). *The Constitution of Society*. Berkeley, CA: University of California Press.

Glaeser, Edward (2011). *Triumph of the City: How our Greatest Invention Makes us Richer, Smarter, Greener, Healthier, and Happier*, New York: Penguin Press.

Gleick, P.H. (2000). "The changing water paradigm: a look at twenty-first century water resources development," *International Water Resources Association*, **25** (1), 127–38.

Gleick, P.H. (2003). "Global freshwater resources: soft-path solutions for the 21st century," *Science*, **302**, 1524–8.

Gleick, P.H., D. Haasz, C. Henges-Jeck, V. Srinivasan, G. Wolff, K. Kao Cushing and A. Mann (2003). "Waste not, want not: the potential for urban water conservation in California," The Pacific Institute, November, available at http://www.colorado.edu/geography/class_homepages/geog_4501_sum14/Western%20Water/J.June%2017/Gleick_waste_not_want_not_full_report.pdf.

Gleick, P.H. and M. Heberger (2012). "The coming mega drought," *Scientific American*, **306**, 1–14.

Gordon, N.D., T.A. McMahon, B.L. Finlayson, C.J. Gippel and R.J. Nathan (2013). *Stream Hydrology: An Introduction for Ecologists*, New York: John Wiley.

Grafton, R. Quentin (2000). "Governance of the commons: a role for the state?," *Land Economics*, **76** (4), 504–17.

Graham, Wade (2013). "Down the drain: how much water goes when it flows?" *Los Angeles Magazine*, 16 September, available at http://www.lamag.com/features-hidden/2013/09/16/down-the-drain-how-much-water-goes-when-it-flows.

Grant, S.B., T.D. Fletcher, D. Feldman, J. Saphores, P.L.M. Cook, M. Stewardson, K. Low, K. Burry and A.J. Hamilton (2013). "Adapting urban water systems to a changing climate: lessons from the Millennium Drought in Southeast Australia," *Environmental Science & Technology*, **47** (19), 10727–34.

Grant, S.B., R.M. Litton-Mueller and J.H. Ahn (2011). "Measuring and modeling the flux of fecal bacteria across the sediment–water interface in a turbulent stream," *Water Resources Research*, **47** (5).

Grant, S., J.D. Saphores, D.L. Feldman, A.J. Hamilton, T.D. Fletcher, P.L.M. Cook, M. Stewardson et al. (2012). "Taking the 'waste' out of 'wastewater' for human water security and ecosystem sustainability," *Science*, **337**, 681–6.

Grebel, J.E., S.K. Mohanty, A.A. Torkelson, A.B. Boehm, C.P. Higgins,

R.M. Maxwell, K.L. Nelson and D.L. Sedlak (2013). "Engineering infiltration systems for urban stormwater reclamation," *Environmental Engineering Science*, **30**, 437–54.

Green, Dorothy (2007). *Managing Water: Avoiding Crisis in California*, Berkeley, CA: University of California Press.

Griffin, D. and K.J. Anchukaitis (2014). "How unusual is the 2012–2014 California drought?," *Geophysical Research Letters*, **41**, 9017–23.

Grimm, N.B., S.H. Faeth, N.E. Golubiewski, C.L. Redman, J. Wu, X. Bai and J.M. Briggs (2008). "Global change and the ecology of cities," *Science*, **319** (5864), 756–60.

Groffman, P.M., A.M. Dorsey and P.M. Mayer (2005). "N processing within geomorphic structures in urban streams," *Journal of the North American Benthological Society*, **24** (3), 613–25.

Groves, D.G., R.J. Lempert, D. Knopman and S.H. Berry (2008). "Preparing for an uncertain future climate in the inland empire – identifying robust water-management strategies," Report – DB-0550-NSF, Santa Monica, CA: RAND Corporation.

Gumprecht, Blake (2001). *The Los Angeles River: Its Life, Death, and Possible Rebirth*, Baltimore, MA: Johns Hopkins University Press.

Gumprecht, Blake (2005). "Who killed the Los Angeles River?," in William Deverell and Greg Hise (eds), *Land of Sunshine: An Environmental History of Metropolitan Los Angeles*, Pittsburgh, PA: University of Pittsburgh, pp. 115–34.

Guo, Y. (2001). "Hydrologic design of urban flood control detention ponds," *Journal of Hydrologic Engineering*, **6** (6), 472–9.

Hamel, P., E. Daly and T.D. Fletcher (2013). "Source-control storm water management for mitigating the impacts of urbanization on baseflow: a review," *Journal of Hydrology*, **485**, 201–11.

Hamel, P., T.D. Fletcher, C.J. Walsh and E. Plessis (2011). "Quantifying the restoration of evapotranspiration and groundwater recharge by vegetated infiltration systems," in Proceedings of the 12th International Conference on Urban Drainage, Porto Alegre, Brazil, 10–16 September.

Hanak, Ellen and Matthew Davis (2009). "Lawns and water demand in California," *California Economic Policy*, **2** (2), 1–22.

Hanak, Ellen, J. Lund, B. Thompson, W. Bowman Cutter, B. Gray, D. Houston, R. Howitt et al. (2012a). "Water and the California economy," Oakland: Public Policy Institute of California.

Hanak, E., J. Lund, A. Dinar, B. Gray, R. Howitt, J. Mount, P. Moyle and B. Thompson (2012b). *Managing California's Water: From Conflict to Reconciliation*, San Francisco, CA: Public Policy Institute of California.

Hatt, B.E., A. Deletic and T.D. Fletcher (2006). "Integrated treatment and recycling of storm water: a review of Australian practice," *Journal of Environmental Management*, **79** (1), 102–13.

Hatt, B.E., T.D. Fletcher and A. Deletic (2008). "Hydrologic and pollutant removal performance of stormwater biofiltration systems at the field scale," *Journal of Hydrology*, **365**, 310–21.

Hatt, B.E., T.D. Fletcher, C.J. Walsh and S.L. Taylor (2004). "The influence of urban density and drainage infrastructure on the concentrations and loads of pollutants in small streams," *Environmental Management*, **34** (1), 112–24.

Hayworth, J.S., G. Glonek, E.J. Maynard, P.A. Baghurst and J. Finlay-Jones (2006). "Consumption of untreated tank rainwater and gastroenteritis among young children in South Australia," *International Journal of Epidemiology*, **35**, 1051–58.

Head, Lesley and Pat Muir (2007). "Changing cultures of water in eastern Australian backyard gardens," *Social & Cultural Geography*, **8** (6), 889–905.

Heather, Peter (2006). *The Fall of the Roman Empire: A New History of Rome and the Barbarians*, Oxford: Oxford University Press.

Hirschman, D., K. Collins and T. Schueler (2008). "The runoff reduction method: technical memorandum," Ellicott City, MD: Center for Watershed Protection & Chesapeake Stormwater Network.

Hise, G. and W. Deverell (2005). "Introduction: the metropolitan nature of Los Angeles," in W. Deverell and G. Hise (eds), *Land of Sunshine: An Environmental History of Metropolitan Los Angeles*, Pittsburgh, PA: University of Pittsburgh Press, pp. 1–12.

Hispagua: Sistema Español de Información sobre el Agua (2009). "Trasvases en América: México," available at http://hispagua.cedex.es/sites/default/files/especiales/Trasvases/mexico.html.

Hodgins, Maureen (2010). "North American residential water use trends since 1992," *Drinking Water Research*, January–March, **21** (1), 19–20.

Hoffman, J. (2010). "Using the water bill to foster conservation," *On Tap*, Winter, 18–22, available at http://www.nesc.wvu.edu/pdf/dw/pub lications/ontap/magazine/OTWI10_features/water_bill_foster_conserva tion.pdf.

Hood, M.J., J.C. Clausen and G.S. Warner (2007). "Comparison of stormwater lag times for low impact and traditional residential development," *Journal of the American Water Resources Association*, **43** (4), 1036–46.

Hopkins, K.G., N.B. Morse, D.J. Bain, N.D. Bettez, N.B. Grimm, J.L. Morse, M.M. Palta et al. (2015). "Assessment of regional variation in streamflow responses to urbanization and the persistence of physiography," *Environmental Science & Technology*, **49** (5), 2724–32.

Howe, C.A., K. Vairavamoorthy and N.P. van der Steen (2011). *SWITCH:*

Sustainable Water Management in the City of the Future – Findings from the SWITCH project, 2006–2011, the Netherlands: European Commission's 6th Framework Programme and SWITCH Consortium partners.

Hua, Ji Wen (2012). *Water Use and Management in Beijing*, Beijing: Institute of Geographic Sciences and Natural Resources Research.

Hughes, J. Donald (2014). *Environmental Problems of the Greeks and Romans: Ecology in the Ancient Mediterranean*, 2nd edn, Baltimore, MD: Johns Hopkins University Press.

Hughes, R.M., S. Dunham, K.G. Maas-Hebner, J.A. Yeakley, C. Schreck, M. Harte, N. Molina et al. (2014). "A review of urban water body challenges and approaches: (1) rehabilitation and remediation," *Fisheries*, **39** (1), 18–29.

Hundley, Norris Jr (2001). *The Great Thirst: Californians and Water: A History*, Berkeley, CA: University of California Press.

Hundley, Norris Jr (2009). *Water and the West: The Colorado River Compact and the Politics of Water in the American West*, Berkeley and Los Angeles, CA: University of California Press.

Hunt, T. (2015). *Ten Cities that Made an Empire*, London: Penguin.

Hurley, Andrew (1995). *Environmental Inequalities: Class, Race, and Industrial Pollution in Gary, Indiana, 1945–1980*, Chapel Hill, NC: University of North Carolina Press.

IMTA (1987). "Overview of the infrastructure for Cutzamala System" ["Visita al Sistema Cutzamala"], Boletín No. 2, Mexico: Instituto Mexicano de Tecnología del Agua.

Imteaz, M.A., A. Ahsan, J. Naser and A. Rahman (2011). "Reliability analysis of rainwater tanks in Melbourne using daily water balance model," *Resources, Conservation and Recycling*, **56** (1), 80–86.

Ingram, H. and C.R. Oggins (1992). "The public trust doctrine and community values in water," *Natural Resources Journal*, **32**, 515–37.

Inman, D. and P. Jeffrey (2006). "A review of residential water conservation tool performance and influences on implementation effectiveness," *Urban Water Journal*, **3**, 127–43.

International Water Association (2013). "Montreal Declaration on Cities of the Future," available at http://www.iwa-network.org/programs/cities-of-the-future/.

IPCC (2007). "Climate change 2007: impacts, adaptation and vulnerability," Contribution of Working Group II to the Fourth Assessment Report of the Intergovernmental Panel on Climate Change, M.L. Parry, O.F. Canziani, J.P. Palutikof, P.J. van der Linden and C.E. Hanson (eds), Cambridge and New York: Cambridge University Press.

IPCC (2014). "Climate change 2014: synthesis report," Contribution of Working Groups I, II and III to the Fifth Assessment Report of the Intergovernmental Panel on Climate Change, core writing team, R.K. Pachauri and L.A. Meyer (eds), Geneva: IPCC.

Jackson, R.B., S.R. Carpenter, C.N. Dahm, D.M. McKnight, R.J. Naiman, S.L. Postel and S.W. Running (2001). "Water in a changing world," *Ecological Applications*, **11**, 1027–45.

Jacob, K., V. Gornitz and C. Rosenzweig (2007). "Vulnerability of the New York City metropolitan area to coastal hazards, including sea level rise: inferences for urban coastal risk management and adaptation policies," in L. McFadden, R.J. Nicholls and E.E. Penning-Rowsell (eds), *Managing Coastal Vulnerability*, Amsterdam and Oxford: Elsevier, pp. 141–58.

Jacobson, Carol R. (2011). "Identification and quantification of the hydrological impacts of imperviousness in urban catchments: a review," *Journal of Environmental Management*, **92** (6), 1438–48.

Jansen, G. (2000). "Urban water transport and distribution," in O. Wikander (ed.), *Handbook of Ancient Water Technology*, Leiden: Brill, pp. 103–25.

Jenerette, G.D., Wanli Wu, Susan Goldsmith, W.A. Marussich and W. John Roach (2006). "Contrasting water footprints of cities in China and the United States," *Ecological Economics*, **57**, 346–58.

Jepperson, R. (1991). "Institutions, institutional effects, and institutionalization," in W.W. Powell and P.J. DiMaggio (eds), *The New Institutionalism in Organizational Analysis*, Chicago, IL: University of Chicago Press, pp. 204–231.

Johnson, E.J., S. Bellman and G.L. Lohse (2003). "Cognitive lock-in and the power law of practice," *Journal of Marketing*, **67**, 62–75.

Kahinda, J.M., A.E. Taigbenu and J.R. Boroto (2007). "Domestic rainwater harvesting to improve water supply in rural South Africa," *Physics and Chemistry of the Earth*, **32** (15), 1050–57.

Kahrl, William M. (1982). *Water and Power: The Conflict over Los Angeles' Water Supply in the Owens Valley*, Berkeley, CA: University of California Press.

Kaika, Maria (2005). *City of Flows: Modernity, Nature, and the City*, New York and London: Routledge.

Kamash, Z. (2012). "An exploration of the relationship between shifting power, changing behaviour and new water technologies in the Roman Near East," *Water History*, **4**, 79–93.

Kamieniecki, S. and A. Below (2009). "Ethical issues in storm water policy implementation: disparities in financial burdens and overall benefits," in John M. Whiteley, Helen Ingram and Richard Perry (eds), *Water, Place, and Equity*, Cambridge, MA: MIT Press, pp. 69–94.

Kanazawa, Mark (2015). *Golden Rules: The Origins of California Water Law in the Gold Rush*, Chicago, IL: University of Chicago Press.

Kelly, V.R., G.M. Lovett, K.C. Weathers, S.E.G. Findlay, D.L. Strayer, D.J. Burns and G.E. Likens (2008). "Long-term sodium chloride retention in a rural watershed: legacy effects of road salt on stream water concentration," *Environmental Science & Technology*, **42** (2), 410–15.

Keremane, G., J. McKay and Z. Wu (2011). "No stormwater in my teacup: an internet survey of residents in three Australian cities," *Water*, April, 118–124.

Kibel, P.S. (2007). *Rivertown: Rethinking Urban Rivers*, Cambridge, MA: MIT Press.

Kingdom, B., R. Liemberger and P. Marin (2006). "The challenge of reducing non-revenue water (NRW) in developing countries," Water Supply and Sanitation Board Discussion Paper Series, Paper No. 8, World Bank, Washington, DC.

Kiparsky, M., D.L. Sedlak, B.H. Thompson and B. Truffer (2013). "The innovation deficit in urban water: the need for an integrated perspective on institutions, organizations, and technology," *Environmental Engineering Science*, **30** (8), 395–408.

Kloss, C. (2008). *Managing Wet Weather with Green Infrastructure Municipal Handbook: Rainwater Harvesting Policies*, Washington, DC: U.S. Environmental Protection Agency.

Koeppel, Gerard T. (2000). *Water for Gotham: A History*, Princeton, NJ: Princeton University Press.

Koo-Oshima, S. and V. Narain (2012). "The hydro-social contract in urban water management in the USA and India," in *Proceedings of the Resilient Cities 2012 Congress*, 3rd Global Forum on Urban Resilience & Adaptation, available at http://resilient-cities.iclei.org/file admin/sites/resilient-cities/files/Resilient_Cities_2012/Digital_Congress_Proceedings/RC2012_Koo-Oshima.pdf.

LA Times (2015). "Unintended consequences of conserving water: leaky pipes, less revenue, bad odors," 1 September, available at http://www.la times.com/local/california/la-me-drought-consequences-20150901-story. html.

Larson, K.L., D. Casagrande, S.L. Harlan and S.T. Yabiku (2009). "Residents' yard choices and rationales in a desert city: social priorities, ecological impacts, and decision tradeoffs," *Environmental Management*, **44** (5), 921–37.

Lauer, S. (2008). "Making storm-water a resource, not a problem," *The California Runoff Rundown – a Newsletter of the Water Education Foundation*, Fall: pp. 1, 4–9, 12.

Lebel, L., J.M. Anderies, B. Campbell, C. Folke, S. Hatfield-Dodds,

T.P. Hughes and J. Wilson (2006). "Governance and the capacity to manage resilience in regional social-ecological systems," *Ecology and Society*, **11**, 19–38.

Lenski, Noel (2002). *Failure of Empire: Valens and the Roman State in the Fourth Century A.D*, Berkeley, CA: University of California Press.

Leung, R.W.K., D.C.H. Li, W.K. Yu, H.K. Chui, T.O. Lee, M.C.M. van Loosdrecht and G.H. Chen (2012). "Integration of seawater and grey water reuse to maximize alternative water resource for coastal areas: the case of the Hong Kong International Airport," *Water Science & Technology*, **65** (3), 410–17.

Leverenz, H.L., G. Tchobanoglous and T. Asano (2011). "Direct potable reuse: a future imperative," *Journal of Water Reuse and Desalination*, **1**(1), 2–10.

Li, H., L.J. Sharkey, W.F. Hunt and A.P. Davis (2009). "Mitigation of impervious surface hydrology using bioretention in North Carolina and Maryland," *Journal of Hydrologic Engineering*, **14** (4), 407–15.

Lim, K.Y., A.J. Hamilton and S.C. Jiang (2015). "Assessment of public health risk associated with viral contamination of harvested urban stormwater for domestic applications," *Science of the Total Environment*, **523**, 95–108.

Lim, K.Y. and S.C. Jiang (2013). "Reevaluation of health risk benchmark for sustainable water practice through risk analysis of rooftop-harvested rainwater," *Water Research*, **47** (20), 7273–86.

Linder, Michael (2006). "Water wars: the battle for Owens Valley" *A KNX Exclusive Investigative Report*, 21 September, available at https://muck rack.com/michaellinder/portfolio/WZ/water-wars-the-battle-for-owens-valley.

Linton, J. (2010). *What is Water? The History of a Modern Abstraction*, Vancouver: University of British Columbia Press.

Liu, M., H. Tian, G. Chen, W. Ren, C. Zhang and J. Liu (2008). "Effects of land-use and land-cover change on evapotranspiration and water yield in China during 1900–2001," *Journal of the American Water Resources Association*, **44** (5), 1193–207.

Lockwood, Harold and Steph Smits (2011). *Supporting Rural Water Supply*, Rugby: Practical Action.

Logan, J.R. and H. Molotch (1987). *Urban Fortunes: The Political Economy of Place*, Berkeley: University of California.

Logan, W.B. and V. Muse (1989). *Smithsonian Guide to Historic America: The Deep South*, New York: Stuart, Tabori, and Chang.

Loorbach, D. and J. Rotmans (2006). "Managing transitions for sustainable development," *Environment & Policy*, **44**, 187–206.

Loperfido, J.V., G.B. Noe, S.T. Jarnagin and D.M. Hogan (2014). "Effects of distributed and centralized stormwater best management practices and land cover on urban stream hydrology at the catchment scale," *Journal of Hydrology*, **519**, 2584–95.

Los Angeles Department of Public Works (2012). "LA's sewers," available at http://www.lacitysan.org/lasewers/sewers/about/index.htm.

Los Angeles Department of Water and Power (LADWP) (2010a). "The story of the Los Angeles aqueduct," available at http://wsoweb.ladwp.com/Aqueduct/historyoflaa/.

Los Angeles Department of Water and Power (LADWP) (2010b). "Urban Water Management Plan," available at www.ladwp.com.

Low, Kathleen G., D.L. Feldman, Stanley B. Grant, Andrew J. Hamilton, K. Gan, J-D. Saphores and M. Arora (2015). "Fighting drought with innovation: Melbourne's response to the Millennium Drought in Southeast Australia," *WIREs Water*, **2** (4).

Lund, Jay R., Richard E. Howitt, Josué Medellín-Azuara and Marion W. Jenkins (2009). "Water management lessons for California from statewide hydro-economic modeling," Center for Watershed Sciences, Department of Civil and Environmental Engineering/ Department of Agricultural and Resource Economics, University of California – Davis, June.

Machiwal, D. and M.K. Jha (2009). "Time series analysis of hydrologic data for water resources planning and management: a review," *Journal of Hydrology and Hydromechanics*, **54** (3), 237–57.

Maddaus, L.A. (2001). "Effects of metering on residential water demand," MSc thesis, UC Davis.

Makropoulos, C.K. and D. Butler (2010). "Distributed water infrastructure for sustainable communities," *Water Resources Management*, **24** (11), 2795–816.

Mayer, P.W., K. DiNatale and W.B. DeOreo (2000). *Show Me the Savings: Do New Homes Use Less Water?* AWWA Annual Conference Proceedings, Denver, CO: AWAA.

Mayer, P.W., W.B. Deoreo, E. Towler and D.M. Lewis (2003). "Residential indoor water conservation study: evaluation of high efficiency indoor plumbing fixture retrofits in single-family homes in the East Bay municipal utility district (EDMUD) service area," The United States Environmental Protection Agency report.

Mayer, P.W., W.B. DeOreo, E. Towler and D.M. Lewis (2004). "Tampa Water Department residential water conservation study: the impacts of high efficiency plumbing fixture retrofits in single-family homes," Tampa, FL: Aquacraft Inc.

McCready, M.S. and M.D. Dukes (2011). "Landscape irrigation scheduling efficiency and adequacy by various control technologies," *Agricultural Water Management*, **98**, 697–704.

McElwain, Mary and C. Herschel (1925). *Strategems and Aqueducts of Rome by Sextus Julius Frontinus*, Cambridge: Harvard University Press.

McPherson, E.G. (1990). "Modeling residential landscape water and energy use to evaluate water conservation policies," *Landscape Journal*, **9**, 122–34.

McQuilkin, Geoffrey (2011). "Stream restoration discussions picking up pace: implementation requires answering many, many questions," *Mono Lake Newsletter*, Summer, p. 5.

Medellín-Azuara, J., L. Mendoza-Espinosa, C. Pells and J.R. Lund (2013). "Pre-feasibility assessment of a water fund for the Ensenada Region: infrastructure and stakeholder analyses," June, Center for Watershed Sciences, UC Davis and Nature Conservancy, Davis, CA.

Melbourne Water (2013). "Water plan," accessed 1 September 2014 at http://melbournewater.com.au/aboutus/reportsandpublications/Documents/Melbourne_Water_2013_Water_Plan.pdf.

Melbourne Water (2014a). "What is a stormwater offset?," available at http://www.melbournewater.com.au/Planning-and-building/schemes/about/Pages/What-are-stormwater-quality-offsets.aspx.

Melbourne Water (2014b). "History of our water supply system," available at http://www.melbournewater.com.au/aboutus/historyandheritage/history-of-our-water-supply-system/pages/history-of-our-water-supply-system.aspx.

Meyer, J.L., M.J. Paul and W.K. Taulbee (2005). "Stream ecosystem function in urbanizing landscapes," *Journal of the North American Benthological Society*, **24** (3), 602–12.

Milesi, C., S.W. Running, C.D. Elvidge, J.B. Dietz, B.T. Tuttle and R.R. Nemani (2005). "Mapping and modeling the biogeochemical cycling of turf grasses in the United States," *Environmental Management*, **36** (3), 426–38.

Miller, G.W. (2006). "Integrated concepts in water reuse: managing global water needs," *Desalination*, **187**, 65–75.

Miller, J.D., H. Kim, T.R. Kjeldsen, J. Packman, S. Grebby and R. Dearden (2014). "Assessing the impact of urbanization on storm runoff in a peri-urban catchment using historical change in impervious cover," *Journal of Hydrology*, **515**, 59–70.

Miller, M.A., B.A. Byrne, S.S. Jang, E.M. Dodd, E. Dorfmeier, M.D. Harris, J. Ames, J. et al. (2010). "Enteric bacterial pathogen detection in southern sea otters (*Enhydra lutris nereis*) is associated with coastal urbanization and freshwater runoff," *Veterinary Research*, **41** (1), 1–13.

Mills, William R. Jr, Susan Bradford, Martin Rigby and Michael Wehner (1998). "Groundwater recharge at the Orange County water district," in

Takashi Asano (ed.), *Wastewater Reclamation and Reuse*, Water Quality Management library, vol. 10, Boca Raton, FL: CRC Press, pp. 1105–36.

Ministerial Advisory Council for the Living Melbourne, Living Victoria Plan for Water (2011). "Living Melbourne, Living Victoria roadmap," March, accessed 1 September 2014 at http://www.depi.vic.gov. au/__data/assets/pdf_file/0009/176472/3770_DSE_Living_Victoria_Road map_1.3MG.pdf.

Minkler, Dana, M., V.B. Vasquez and A.C. Baden (2006). "Community-based participatory research as a tool for policy change: a case study of the Southern California environmental justice collaborative," *Review of Policy Research*, **23** (2), 339–53.

Mitchell, K., N. Wimbush, C. Harty, G. Lampy and G. Sharpley (2008a). "Victorian desalination project environment effects statement report of the inquiry to the Minister for Planning," accessed 3 July 2009 at http://www.dpi.vic.gov.au.

Mitchell, V.G., H.A. Cleugh, C.S.B. Grimmond and J. Xu (2008b). "Linking urban water balance and energy balance models to analyse urban design options," *Hydrological Processes*, **22**, 2891–900.

Mohadjer, J. and D.L. Rice (2004). *Water Conservation Annual Report*, West Jordan: Jordan Valley Water Conservancy District.

Molotch, H. (1976). "The city as a growth machine: toward a political economy of place," *American Journal of Sociology*, **82**, 309–30.

Monash University (2008). "Submission to the Victorian Environment and Natural Resources Committee: inquiry into Melbourne's future water supply," Melbourne: Monash University.

Mono Lake Newsletter (2014). "Celebrating Mono Basin Stream Restoration Agreement," Special Report for Mono Lake Committee Members, March.

Morin, Monte (2012). "Some climate scientists, in a shift, link weather to global warming," *Los Angeles Times*, 12 October.

Morse, C.C., A.D. Huryn and C. Cronan (2003). "Impervious surface area as a predictor of the effects of urbanization on stream insect communities in Maine, USA," *Environmental Monitoring and Assessment*, **89** (1), 95–127.

Moy, Candace (2012). "Rainwater tank households: water savers or water users?," *Geographical Research*, **50** (2), 204–16.

Mueller, E.C. and T.A. Day (2005). "The effect of urban ground cover on microclimate, growth and leaf gas exchange of oleander in Phoenix, Arizona," *International Journal of Biometeorology*, **49**, 244–55.

Mulholland, Catherine (2002). *William Mulholland and the Rise of Los Angeles*, Berkeley, CA: University of California Press.

Murray, C.G. and A.J. Hamilton (2010). "Perspectives on wastewater treatment wetlands and waterbird conservation," *Journal of Applied Ecology*, **47**, 976–85.

Murray, K.B. and G. Häubl (2007). "Explaining cognitive lock-in: the roll of skill-based habits of use in consumer choice," *Journal of Consumer Research*, **34**, 77–88.

Museu d'Historia de Barcelona (2012). "Water/Barcelona: a guide to urban history," available at http://museuhistoria.bcn.cat/sites/default/files/guiaaiguabcn.478_0.pdf.

Muthukumaran, S., K. Baskaran and N. Sexton (2011). "Quantification of potable water savings by residential water conservation and reuse: a case study," *Resources, Conservation and Recycling*, **55** (11), 945–52.

National Research Council (2012). *Water Reuse: Expanding the Nation's Water Supply Through Reuse of Municipal Wastewater*, Washington, DC: The National Academies Press.

Nellor, M.H. and R. Larson (2010). "Assessment of approaches to achieve nationally consistent reclaimed water standards," Alexandria, VA: Water Reuse Research Foundation.

New York City (2011). "History of New York City's water supply system," available at http://www.nyc.gov/html/dep/html/drinking_water/history.shtml.

New York City Department of Environmental Protection (2010). *New York City 2010 Drinking Water Supply and Quality Report*, Flushing, NY: NY DEP.

New York State Department of Environmental Conservation (2010a). "New York City watershed program," available at http://www.dec.ny.gov/land58597.html.

New York State Department of Environmental Conservation (2010b). "Facts about the New York City watershed," available at http://www.dec.ny.gov/lands/58524.html.

Nicholson, N., S.E. Clark, B.V. Long, J. Spicher and K.A. Steele (2009). "Rainwater harvesting for non-potable use in gardens: a comparison of runoff water quality from green vs. traditional roofs," in *Proceedings of World Environmental and Water Resources* Congress, Kansas City, MO, 17–21 May, Reston, VA: ASCE.

Nieswiadomy, M.L. (1992). "Estimating urban residential water demand: effects of price structure, conservation, and education," *Water Resources Research*, **28** (3), 609–15.

Nieswiadomy, M.L. and D.J. Molina (1989). "Comparing residential water demand estimates under decreasing and increasing block rates using household data," *Land Economics*, **65** (3), 280–89.

Novotny, Vladimir (2010). "Water and energy footprints for sustainable communities," *Proceedings of the Singapore International Water Week*

Conference, 28 June–2 July, available at http://aquanovallc.com/wp-content/uploads/2010/12/Singapore-2010.pdf.

Nowak, D.J., D.E. Crane and J.C. Stevens (2006). "Air pollution removal by urban trees and shrubs in the United States," *Urban Forestry & Urban Greening*, **4**, 115–23.

Nowak, D. and J. Dwyer (2007). "Understanding the benefits and costs of urban forest ecosystems," in J.E. Kuser (ed.), *Urban and Community Forestry in the Northeast*, Dordrecht: Springer, pp. 25–46.

OCWD/OCSD Partnership (2004). "The OCWD/OCSD partnership," Orange County Water District, Fountain Valley, CA, available at http://www.ocwd.com/gwrs/the-ocwdocsd-partnership/.

O'Driscoll, M., S. Clinton, A. Jefferson, A. Manda and S. McMillan (2010). "Urbanization effects on watershed hydrology and in-stream processes in the southern United States," *Water*, **2** (3), 605–48.

Office of Living Victoria (2014). Melbourne, Victoria, available at http://www.livingvictoria.vic.gov.au/index.html.

Office of the Premier (2007). "Desalination and pipelines to secure water supplies," media release, Department of Premier and Cabinet, 19 June, accessed 24 August 2009 at http://www.dpc.vic.go.au.

Olivera, Marcela and Jorge Viana (2003). "Winning the water war," *Human Rights Dialogue*, Spring, pp. 10–11.

Ometo, J.P.H.B., L.A. Martinelli, M.V. Ballester, A. Gessner, A.V. Krusche, R.L. Victoria and M. Williams (2000). "Effects of land use on water chemistry and macroinvertebrates in two streams of the Piracicaba river basin, south-east Brazil, *Freshwater Biology*, **44** (2), 327–37.

Orange County Community Indicators Project (2011). "Orange County 2011 community indicators," available at http://ocgov.com/civicax/file bank/blobdload.aspx?BlobID=4098.

Ostrega, S.F. (1996). "New York City: where conservation, rate relief and environmental policy meet," report, New York Department of Environmental Protection, Bureau of Water and Energy Conservation.

Ostrom, E. (2010). "Beyond markets and states: polycentric governance of complex economic systems," *American Economic Review*, **100** (June), 1–33.

Otaki, Y., M. Otaki and O. Sakura (2007). "Water systems and urban sanitation: a historical comparison of Tokyo and Singapore," *Journal of Water and Health*, **5** (2), 259–65.

Padowski, Julie C., Steven M. Gorelick, Barton H. Thompson, Scott Rozelle and Scott Fendorf (2015). "Assessment of human–natural system characteristics influencing global freshwater supply vulnerability," *Environmental Research Letters*, **10**, 104014.

Pahl-Wostl, C. (2005). Information, public empowerment, and the management of urban watersheds, *Environmental Modelling & Software*, **20**, 457–67.

Pahl-Wostl, C., J. Sendzimir, P. Jeffrey, J. Aerts, G. Berkamp and K. Cross (2007). "Managing change toward adaptive water management through social learning," *Ecology & Society*, **12** (2).

Parolari, Anthony J., Gabriel G. Katul and Amilcare Porporato (2015). "The Doomsday Equation and 50 years beyond: new perspectives on the human-water system," *WIREs Water*, **2** (4), 407–14.

Pellow, David N. (2002). *Garbage Wars: The Struggle for Environmental Justice in Chicago*, Cambridge, MA: MIT Press.

Perreault, Thomas (2005). "State restructuring and the scale politics of rural water governance in Bolivia," *Environment and Planning A*, **37** (2), 263–84.

Persson, J., N.L.G. Somes and T.H.F. Wong (1999). "Hydraulics efficiency of constructed wetlands and ponds," *Water Science & Technology*, **40** (3), 291–300.

Petrucci, G., E. Rioust, J. Deroubaix and B. Tassin (2013). "Do storm water source control policies deliver the right hydrologic outcomes?" *Journal of Hydrology*, **485**, 188–200.

Petrucci, G., F. Rodriguez, J.F. Deroubaix and B. Tassin (2014). "Linking the management of urban watersheds with the impacts on the receiving water bodies: the use of flow duration curves," *Water Science & Technology*, **70** (1), 127–35.

Pimentel, David, Bonnie Berger, David Filiberto and Michelle Newton (2004). "Water resources: agricultural and environmental issues," *BioScience*, **54** (10), 909–18.

Planning Institute Australia (2014). "Water and planning," available at http://www.planning.org.au/policy/water-and-planning.

Po, M., B.E. Nancarrow, Z. Leviston, N.B. Porter, G.J. Syme and J.D. Kaercher (2005). "Predicting community behaviour in relation to wastewater reuse: what drives decisions to accept or reject?," Water for a Healthy Country: National Research Flagships, Perth: CSIRO Land and Water.

Poff, N.L. and J.K.H. Zimmerman (2010). "Ecological responses to altered flow regimes: a literature review to inform the science and management of environmental flows," *Freshwater Biology*, **55** (1), 194–205.

Pomeroy, Earl (1965). *The Pacific Slope: A History of California, Oregon, Washington, Idaho, Utah, and Nevada*, New York: Alfred A. Knopf.

Post, Allison (2009). "The paradoxical politics of water metering in Argentina," in *Poverty in Focus*, International Policy Centre for Inclusive Growth, **18**, August, Bureau for Development Policy, UNDP, pp. 16–18.

Potts, J. (2009). "The innovation deficit in public services: the curious problem of too much efficiency and not enough waste and failure," *Innovation: Management, Policy & Practice*, **11**, 34–43.

Proenca, L.C., E. Ghisi, D.D.F. Tavares and G.M. Coelho (2011). "Potential for electricity savings by reducing potable water consumption in a city scale," *Resources, Conservation and Recycling*, **55**, 960–65.

Purcell, Nicholas (1994). "The arts of government," in John Boardman, Jasper Griffin and Oswyn Murray (eds), *The Roman World*, Oxford: Oxford University Press, pp. 150–81.

Reeves, R.L., S.B. Grant, R.D. Mrse, C.M.C. Oancea, B.F. Sanders and A.B. Boehm (2004). "Scaling and management of fecal indicator bacteria in runoff from a coastal urban watershed in Southern California," *Environmental Science & Technology*, **38**, 2637–48.

Reheis, M.C. (1997). "Dust deposition downwind of Owens (dry) Lake: 1991–1994: preliminary findings," *Journal of Geophysical Research*, **102**, 25999–6008.

Reichold, L., E.M. Zechman, E.D. Brill and H. Holmes (2010). "Simulation optimization framework to support sustainable watershed development by mimicking the predevelopment flow regime," *Water Research*, **136** (3), 366–75.

Reilly, J.F., A.J. Horne and C.D. Miller (1999). "Nitrate removal from a drinking water supply with large free surface constructed wetlands prior to groundwater recharge," *Ecological Engineering*, **14**, 33–47.

Renn, O. and M.C. Roco (2006). "White paper on nanotechnology risk governance," Geneva: International Risk Governance Council.

Rippy, M.A., R. Stein, B.F. Sanders, K. Davis, K. McLaughlin, J.F. Skinner, J. Kappeler and S.B. Grant (2014). "Small drains, big problems: the impact of dry weather runoff on shoreline water quality at enclosed beaches," *Environmental Science & Technology*, **48**, 14168–77.

Rogers, E.M. (1983). *Diffusion of Innovations*, New York: Free Press.

Rose, S. and N.E. Peters (2001). "Effects of urbanization on streamflow in the Atlanta area (Georgia, USA): a comparative hydrological approach," *Hydrological Processes*, **15** (8), 1441–57.

Rosenzweig, C. and W.D. Solecki (eds) (2001). "Climate change and a global city: the potential consequences of climate variability and change – Metro East Coast," New York: Columbia Earth Institute, Columbia University, available at http://metroeast_climate.ciesin.columbia.edu/.

Rosenzweig, C., D.C. Major, K. Demong, C. Stanton, R. Horton and M. Stults (2007). "Managing climate change risks in New York City's water system: assessment and adaptation planning," *Mitigation and Adaptation Strategies for Global Change*, **12** (8), 1391–409.

Rotmans, J., R. Kemp and M. van Asselt (2001). "More evolution than revolution: transition management in public policy," *Foresight*, **3** (1), 15–31.

Rousseau, D.P.L., E. Lesage, A. Story, P.A. Vanrolleghem and N. De Pauw (2008). "Constructed wetlands for water reclamation," *Desalination*, **218** (1–3), 181–9.

Roy, A.H., L.K. Rhea, A.L. Mayer, W.D. Shuster, J.J. Beaulieu, M.E. Hopton, M.A. Morrison and A.S. Amand (2014). "How much is enough? Minimal responses of water quality and stream biota to partial retrofit storm water management in a suburban neighborhood," *PLoS One*, **9** (1), e85011.

Roy, A.H., S.J. Wenger, T.D. Fletcher, C.J. Walsh, A.R. Ladson, W.D. Shuster, H.W. Thurston, and R.R. Brown (2008). "Impediments and solutions to sustainable, watershed-scale urban storm water management: lessons from Australia and the United States," *Environmental Management*, **42**, 344–59.

Saha, D. and R.G. Paterson (2008). "Local government efforts to promote the 'three As' of sustainable development: survey in medium to large cities in the U.S.," *Journal of Planning Education and Research*, **28**, 21–37.

Saliba, C. and K. Gan (2012). "Energy density maps in water demand management," Report No. E6109 to the Yarra Valley Water District, Melbourne: Yarra Valley Water.

Satterthwaite, D, (2000). "Will most people live in cities?," *British Medical Journal*, **321** (7269), 1143–5.

SCCWRP (Southern California Coastal Water Research Project) (2010). *Project Area: Dry-Weather Runoff Pollutant Loading*, Costa Mesa, CA: SCCWRP.

Schnaiberg, A. and K.A. Gould (1994). *Environment and Society: The Enduring Conflict*, New York: St Martin's.

Schnoor, J.L. (2009). "NEWater future?," *Environmental Science & Technology*, **43**, 6441–2.

Schroeder, E., G. Tchobanoglous, H.L. Leverenz and T. Asano (2012). *Direct Potable Reuse: Benefits for Public Water Supplies, Agriculture, the Environment, and Energy Conservation*, Fountain Valley, CA: National Water Research Institute.

Scott, W.R. (2013). *Institutions and Organizations: Ideas, Interests, and Identities*, 4th edn, Thousand Oaks, CA: Sage.

Selbig, W.R. and R.T. Bannerman (2008). *A Comparison of Runoff Quantity and Quality from Two Small Basins Undergoing Implementation of Conventional and Low-Impact-Development (LID) Strategies: Cross Plains, Wisconsin, Water Years 1999–2005*, U.S. Geological Survey report 2008–5008, Washington, DC: U.S. Geological Survey.

Sessions, George (1995). *Deep Ecology for the 21st Century*, Boston, MA: Shambhala Publications.

Shaughnessy, Edward L. (2000). *China: Empire and Civilization*, Oxford: Oxford University Press.

Shove, Elizabeth (2003a). *Comfort, Cleanliness and Convenience*, Oxford: Berg.

Shove, Elizabeth (2003b). "Converging conventions of comfort, cleanliness and convenience," *Journal of Consumer Policy*, **26** (4), 395–418.

Shuster, W.D., J. Bonta, H. Thurston, E. Warnemuende and D.R. Smith (2005). "Impacts of impervious surface on watershed hydrology: a review," *Urban Water Journal*, **2** (4), 263–75.

Shuster, W. and L. Rhea (2013). "Catchment-scale hydrologic implications of parcel-level storm water management (Ohio USA)," *Journal of Hydrology*, **485**, 177–87.

Siriwardene, N., M. Quilliam and P. Roberts (2011). "How effective is Target 155 in Melbourne? Insight from climate correction modelling," paper presented at the 4th AWA National Water Efficiency Conference, Melbourne.

Smart Water Fund (2002). "Smart water fund changes," City West Water, South East Water, Yarra Valley Water, Melbourne Water and the Department of Environment and Primary Industries, available at http://www.smartwater.com.au.

Smith, A. and A. Stirling (2010). "The politics of social-ecological resilience and sustainable sociotechnical transitions," *Ecology and Society*, **15**, 11–23.

Sofoulis, Zoë (2005). "Big water, everyday water: a sociotechnical perspective," *Continuum: Journal of Media & Cultural Studies*, **19** (4), 445–63.

Sofoulis, Zoë (2006). "Changing water cultures," in E. Probyn, S. Muecke and A. Shoemaker (eds), *Creating Value: The Humanities and Their Publics*, Canberra: Australian Academy of the Humanities, pp. 105–15.

Sofoulis, Zoë and Carolyn Williams (2008). "From pushing atoms to growing networks: cultural innovation and co-evolution in urban water conservation," *Social Alternatives*, **27** (3), 50–57.

Srinivasan, V., S.M. Gorelick and L. Goulder (2010). "Sustainable urban water supply in south India: Desalination, efficiency improvement, or rainwater harvesting?," *Water Resources Research*, **46**, W10504.

Starr, K. (1985). *Inventing the Dream: California through the Progressive Era*, New York: Oxford University Press.

Stephens and Associates Inc. (2013). "Hydrologic characterization and water balance development, Newport Bay Watershed, Swamp of the Frogs, Orange County, California," technical report prepared for Orange

County Public Works, Albuquerque, NM: Daniel B. Stephens and Associates Inc.

Stevens, T.H., J. Miller and C. Willis (1992). "Effect of price structure on residential water demand," *Journal of the American Water Resources Association*, **28**, 681–5.

Stokes, J.R. and A. Horvath,(2009). "Energy and air emission effects of water supply," *Environmental Science & Technology*, **43** (8), 2680–87.

Sule, S. (2003). "Understanding our civic issues: Mumbai's water supply," Mumbai: The Bombay Community Public Trust.

Supski, Sian and Jo Lindsay (2013). *Australian Domestic Water Use Cultures: A Literature Review*, Melbourne: Cooperative Research Centre for Water Sensitive Cities.

Surbeck, C.Q., S.C. Jiang and S.B. Grant (2010). "Ecological control of fecal indicator bacteria in an urban stream," *Environmental Science & Technology*, **44**, 631–7.

Swyngedouw, Eric (2007). "Water, money, and power," *Socialist Register*, **43**, 195–212.

Taylor, S.L., S.C. Roberts, C.J. Walsh and B.E. Hatt (2004). "Catchment urbanization and increased benthic algal biomass in streams: linking mechanisms to management," *Freshwater Biology*, **49** (6), 835–51.

The Age (2014). "Office of Living Victoria killed off by new government," 12 December, available at http://www.theage.com.au/victoria/office-of-living-victoria-killed-off-by-new-government-20141212-125vmd.html.

Tokyo Waterworks Bureau (2013). "Outline of the Tokyo Waterworks Bureau," available at http://www.waterprofessionals.metro.tokyo.jp/pdf/wst_02.pdf.

Tortajada, C. (2006). "Who has access to water? Case study of Mexico City Metropolitan Area," Human Development Report Office Occasional Paper, available at http://www.unwater.org/fileadmin/templates/unwater/unwater_new/images/teaser.png.

Tortajada, C. and E. Casteian (2003). "Water management for a megacity: Mexico City Metropolitan Area," *Ambio*, **32** (2), 124–9.

Townsend-Small, A. and C.I. Czimczik (2010). "Carbon sequestration and greenhouse gas emissions in urban turf," *Geophysical Research Letters*, **37** (2).

Townsend-Small, A., D.E. Pataki, H. Liu, Z. Li, Q. Wu and B. Thomas (2013). "Increasing summer river discharge in southern California, USA, linked to urbanization," *Geophysical Research Letters*, **40** (17), 4643–7.

Turner, A., S. White, K. Beatty and A. Gregory (2004). "Results of the largest residential demand management program in Australia," Study Report, Sydney: Institute for Sustainable Futures for Sydney Water Corporation.

Turner, B.L. and Jeremy A. Sabloff (2012). "Classic period collapse of the Central Maya Lowlands: insights about human–environment relationships for sustainability," *PNAS*, **109** (35), 13908–914.

UN Water (2014). The United Nations Inter-agency Mechanism on all Freshwater Related Issues. http://www.unwater.org/statistics/statistics-detail/en/c/246663/.

UNESCO-IHE – Institute for Water Education (2011). *Water Solutions – UNESCO-IHE in Partnership*, Delft: UNESCO.

United Nations Human Settlements Programme (2011). *Cities and Climate Change: Global Report on Human Settlements, 2011*, London: Earthscan.

U.S. Department of the Interior, Bureau of Reclamation (USBR) (2008). "Summary of smart controller water savings studies: literature review of water savings studies for weather and soil moisture based landscape irrigation control devices," Final Technical Memorandum No. 86-68210-SCAO-01.

U.S. Environmental Protection Agency (U.S. EPA) (1972). Clean Water Act, 33 U.S.C. §1251 et seq.

U.S. Environmental Protection Agency (U.S. EPA) (2004). *Guidelines for Water Reuse*, Washington, DC. EPA/624/R-04/108.

U.S. Environmental Protection Agency (U.S. EPA) (2012). "Chapter 3 – Development and Implementation of The TMDL – Guidance for Water Quality-Based Decisions: The TMDL Process," March 2012 update.

U.S. Environmental Protection Agency (U.S. EPA) (2016). "National Pollutant Discharge Elimination System (NPDES)," available at http://cfpub.epa.gov/npdes/.

Utz, R.M., K.N. Eshleman and R.H. Hilderbrand (2011). "Variation in physicochemical responses to urbanization in streams between two Mid-Atlantic physiographic regions," *Ecological Applications*, **21**, 402–15.

Van de Meene, S., R.R. Brown and M.A. Farrelly (2011). "Towards understanding governance for sustainable urban water management," *Global Environmental Change*, **21** (3), 1117–27.

Van der Brugge, R. and J. Rotmans (2007). "Towards transition management of European water resources," *Water Resources Management*, **21**, 249–67.

VanderBrug, Brian (2009). "In the Owens Valley, resentment again flows with the water," *Los Angeles Times*, 16 May, B-1.

Varon, M.P. and D. Mara (2004). "Waste stabilisation ponds," IRC International Water and Sanitation Centre.

Vietz, G.J., M.J. Sammonds, C.J. Walsh, T.D. Fletcher, I.D. Rutherfurd and M.J. Stewardson (2014). "Ecologically relevant geomorphic attributes of streams are impaired by even low levels of watershed effective imperviousness," *Geomorphology*, **206**, 67–78.

Village of Croton (2010). "History of the New Croton Dam," available at http://village.croton-on-hudson.ny.us/public_documents/croton hudsonny_webdocs/historicalsociety/crotondam.

Vorosmarty, C.J., P. Green, J. Salisbury and R.B. Lammers (2000). "Global water resources: vulnerability from climate change and population growth," *Science*, **289** (5477), 284–8.

Victoria Water Management Strategy (VWMS) (1994). Catchment and Land Protection Act 1994 (the CaLP Act), available at http://delwp. vic.gov.au/water/governing-water-resources/catchment-management-au thorities#sthash.9pJN4HVS.dpuf.

Walsh, Chris (2007). "The keys to restoring the Yarra River," paper presented at the seminar "Water quality in the Yarra River," Port Phillip and Westernport Catchment Management Authority and Yarra Riverkeepers Association, Melbourne, 15 June.

Walsh, C.J., T.D. Fletcher and M.J. Burns (2012). "Urban storm water runoff: a new class of environmental flow problem," *PLoS One*, **7** (9), e45814.

Walsh, C.J., T.D. Fletcher and A.R. Ladson (2005b). "Stream restoration in urban catchments through redesigning storm water systems: looking to the catchment to save the stream," *Journal of the North American Benthological Society*, **24**, 690–705.

Walsh, C.J. and J. Kunapo (2009). "The importance of upland flow paths in determining urban effects on stream ecosystems," *Journal of the North American Benthological Society*, **28** (4), 977–90.

Walsh, C.J., A.H. Roy, J.W. Feminella, P.D. Cottingham, P.M. Groffman and R.P. Morgan (2005a). "The urban stream syndrome: current knowledge and the search for a cure," *Journal of the North American Benthological Society*, **24** (3), 706–23.

Walton, J. (1993). *Western Times and Water Wars: State, Culture, and Rebellion in California*, Berkeley, CA: University of California Press.

Wang, U. (2014). "New technology tools aim to reduce water use," *The Wall Street Journal*, 18 May.

Water Education Foundation (2013). *Layperson's Guide to Integrated Regional Water Management*, Sacramento, CA: Water Education Foundation.

Water Innovation Centre (2013). "Governance and management," International Institute for Sustainable Development, available at http://www.iisd.org/wic/research/governance/.

Watson, F., R. Vertessy, T. McMahon, B. Rhodes and I. Watson (2001). "Improved methods to assess water yield changes from paired-catchment studies: application to the Maroondah catchments," *Forest Ecology and Management*, **143** (1), 189–204.

Welty, C., L. Band, R.T. Bannerman, D.B. Booth, R.R. Horner, C.R. O'Melia, R.E. Pitt et al. (2009). *Urban Stormwater Management in the United States*, Water Science and Technology Board, Washington, DC: National Research Council.

Wenger, S.J., A.H. Roy, C.R. Jackson, E.S. Bernhardt, T.L. Carter, S. Filoso and C.A. Gibson (2009). "Twenty-six key research questions in urban stream ecology: an assessment of the state of the science," *Journal of the North American Benthological Society*, **28** (4), 1080–98.

Westchester County Department of Planning (2009). *The Croton Plan for Westchester: The Comprehensive Croton Watershed Water Quality Protection Plan*, September, available at www.westchestergov com crotonplan.

Willensky, Elliot and Norval White (1988). *American Institute of Architects Guide to New York City*, 3rd edn, Orlando, FL: Harcourt Brace.

Williams, W.D. (2001). "Anthropogenic salinization of inland waters," *Hydrobiologia*, **466**, 329–37.

Willoughby, L. 1999. *Flowing through Time: A History of the Lower Chattahoochee River*, Tuscaloosa, AL: University of Alabama Press.

Wong, S. (2007). "China bets on massive water transfers to solve crisis," *International Rivers*, 15 December, reproduced from *World Rivers Review*.

Wong, Tony H.F., R. Allen, J. Beringer, R.R. Brown, V. Chaudhri, A. Deletić, T.D. Fletcher et al. (2011). *Blueprint 2011: Stormwater Management in a Water Sensitive City*, Centre for Water Sensitive Cities, University of Melbourne.

World Civil Society Forum (2002). *Strengthening International Cooperation*, Geneva, available at http://www.worldcivilsociety.org/pages/164.

World Health Organization (WHO) (2006). "WHO guidelines for the safe use of wastewater, excreta and greywater," Geneva: World Health Organization.

World Health Organization (2014). "Water-related diseases," 1 June, Geneva: World Health Organization.

World Resources Institute (2012). "*AQUEDUCT – Measuring and mapping water* risk," Washington, DC: World Resources Institute.

Wu, Z., J. McKay and G. Keremane (2012). "Issues affecting community attitudes and intended behaviors in stormwater reuse: A case study of Salisbury, South Australia," *Water*, **4** (4), 835–47.

Yarra Riverkeepers (2015a). "Description," available at http://yarrariver. org.au/?page_id=14.

Yarra Riverkeepers (2015b), "River issues," available at http://yarrariver. org.au/?page_id=18.

Yusuf, K.A. (2007). "Evaluation of groundwater quality characteristics in Lagos City," *Journal of Applied Sciences*, **7** (13), 1780–84.

Zérah, M.H. (2008). "Splintering urbanism in Mumbai: contrasting trends in a multilayered society," *Geoforum*, **39**, 1922–32.

Zhang, L., W.R. Dawes and G.R. Walker (1999). *Predicting the Effect of Vegetation Changes on Catchment Average Water Balance; Technical Report 99/12*, Cooperative Research Centre for Catchment Hydrology: Victoria, Australia.

Zhang, L., W.R. Dawes and G.R. Walker (2001). "Response of mean annual evapotranspiration to vegetation changes at catchment scale," *Water Resources Research*, **37** (3), 701–708.

SELECTED PRIMARY SOURCES ON MELBOURNE

City of Melbourne. Eastern Melbourne Parks and Gardens Stormwater Harvesting Scheme 2013. Accessed 24 September 2014 at http://www.environment.gov.au/node/25191.

City West Water. Altona Recycled Water Project Information Sheet. Accessed 1 September 2014 at http://www.citywestwater.com.au/documents/Altona_RW_Project.pdf.

City West Water, South East Water, Yarra Valley Water, Melbourne Water. Melbourne Joint Water Conservation Plan. Annual Report 2010/2011.

Darling Street Stormwater Harvesting Project. Accessed 1 September 2014 at http://www.clearwater.asn.au/resource-library/case-studies/darling-street-stormwater-harvesting-project.php.

Melbourne Water. 2013 Water Plan. Accessed 1 September 2014 at http://melbournewater.com.au/aboutus/reportsandpublications/Documents/Melbourne_Water_2013_Water_Plan.pdf.

Ministerial Advisory Council for the Living Melbourne, Living Victoria Plan for Water. Living Melbourne, Living Victoria Roadmap 2011, March. Accessed 1 September 2014 at http://www.depi.vic.gov.au/__data/assets/pdf_file/0009/176472/3770_DSE_Living_Victoria_Roadmap_1.3MG.pdf.

Office of Living Victoria. State Government of Victoria. Melbourne's Water Future 2013, July. Accessed 1 September 2014 at http://www.livingvictoria.vic.gov.au/PDFs/Melbourne's_Water_Future_full.pdf.

Official launch of Darling Street Stormwater Harvesting Project. 10 June 2012. Accessed 1 September 2014 at http://www.clearwater.asn.au/news/official-launch-of-darling-street-stormwater-harvesting-project.php.

Southern Rural Water, Melbourne Water, URS. Regional Environment

Improvement Plan Werribee Irrigation District Class A Recycled Water Scheme. Accessed 16 September 2014 at http://www.srw.com.au/Files/Technical_reports/WID_REIP_Final_July_2009.pdf.

State Government of Victoria. Right Water. Accessed 24 September 2014 at http://www.rightwater.vic.gov.au/list.html.

Yarra Valley Water. Wallan Sewage Treatment Plant Fact Sheet. Accessed 1 September 2014 at https://www.yvw.com.au/yvw/groups/public/documents/document/yvw1002064.pdf.

Yarra Valley Water. Whittlesea Sewage Treatment Plant Fact Sheet. Accessed 1 September 2014 at http://www.yvw.com.au/yvw/groups/public/documents/document/yvw1002065.pdf.

Index